Communications Networks

Communications
Networks
Michael F. Hordeski

TAB Professional and Reference Books

Division of TAB BOOKS Inc.
Blue Ridge Summit, PA

FIRST EDITION
FIRST PRINTING

Library of Congress Cataloging in Publication Data

Hordeski, Michael F.
   Communications networks / by Michael F. Hordeski.
      p.   cm.
   Includes index.
   ISBN 0-8306-3188-7 （H）
   1. Computer networks—Management.   2. Computer network
architectures.   I. Title.
TK5105.5.H67   1989
004.6—dc20   89-33597
                                                    CIP

TAB BOOKS Inc. offers software for sale. For information and a catalog, please contact
TAB Software Department, Blue Ridge Summit, PA 17294-0850.

Questions regarding the content of this book should be addressed to:

   Reader Inquiry Branch
   TAB BOOKS Inc.
   Blue Ridge Summit, PA 17294-0214

Acquisitions Editor: Larry Hager
Technical Editor: Steven L. Burwen
Series Design: Jaclyn B. Saunders

Contents

Acknowledgments

A book such as this is never possible without the help of others. I would like to thank such communication professionals as Donald Gladstone and Ellen Hancock of IBM; Ron Cook of E. F. Hutton; Shirley Radack of the NBS Institute for Computer Sciences and Technology; Paul Demko and J. J. Cinecoe of Wang Laboratories; George Colony of Forrester Research; Michael Smith of the University of Wisconsin at Madison; Jerry Mulvenna of NBS; Michael Kaminski of General Motors; Karl Kozarsky of Probe Research; Allen Gersho of the University of California at Santa Barbara; Maren Symonds of Octel; Mike Baker of Baxter Healthcare; Raj Melville of Booz, Allen, and Hamilton; Dan Riley of Clover Electronics; Charles Robins of Rabbit Software; Tom Osborn of Harris; Don Van Doren of Vanguard Telecommunications; Frank Ravest of Northern Telecom; Gus Bender of Travelers; John Lusa of TPT; and Gerry Sawkins of Ungermann-Bass.

Thanks to Larry Hager at TPR for believing in the project and a special thanks to Dee of Jablon Computer Associates for making sense of my very rough notes.

Introduction

Ever since the first computers were shipped to businesses in the early '50s, there has been great pressure to distribute computer intelligence. The motive has generally been better control of the computing resource or more efficiency in information processing.

Today, with so many companies operating network-based applications, there is also another motive: saving time for the professional and administrative staff. The more distributed information processing is, the greater is the productivity of all employees.

The benefits in time saved approach 25 percent for professionals and 30 percent for administrative support staff. A 20-month cost recovery and a return on investment of almost 60 percent can also be projected. A real advantage is the competitive edge achieved.

Network practitioners seek to find higher and higher levels of integration. This even includes the buildings in which automation is taking place.

The challenge of systems integration is harnessing change. What makes the challenge so demanding is the speed with which the changes are taking place. For example, in 1980 there were less than 4 million electronic keyboards in offices in the United States. By 1990 it is predicted that there will be almost 50 million. In 1980, digital telephone switches were rare, digital telephones even rarer. By 1990, almost all switches will be digital and have provisions for switching data traffic. The telephone companies also will be heavily marketing the integrated-services digital networks (ISDNs), which are now in several pilot phases.

Add to the permanent state of fluctuation brought to the market by continual technological advances and sociopolitical changes such as the deregulation of basic industries and the breakup of AT&T and you have even more unknowns.

These *unknowns* make the challenge of systems integration more difficult. It is one thing to link systems into an integrated entity; it is another to do it when system technologies sometimes advance along different paths, but always at lightning fast speeds.

Add to this the technology of the so-called "intelligent" buildings, where tenants share not only heat, light, transportation, environmental controls, and building management (computerized), but also telecommunications switches, cabling, satellite dishes, and even computers themselves.

The building-controls business is no less immune to digitization than other activities, and the result is a general integration of fire protection, security, energy management, and environmental control into a single, computer-based system operating over a single wiring system. Sensor-based, computer-driven control of heating systems, for instance, can cut up to 40 percent from energy bills by turning off thermostats. Potentially more significant are these savings that result from better system operation during emergencies.

The newest PBXs allow enough programming to permit the kind of complex switching and administration needed to allow tenant sharing of telecommunications switching and high-speed lines. Economies of scale are such that unrelated businesses sharing telecommunications services can cut up to 30 percent from their telephone bills.

The new generation of intelligent networks will have both computer-controlled building management and shared tenant communications. In the ultimate scenario, everything is integrated. The network is so intelligent that its users can be imprecise in making demands on it, and interfaces will be so standardized that anything can talk to anything. We are not there yet. It is tough enough understanding today's technology and systems that integrate only a portion at a time. Successive waves of integration are coming rapidly. They could present us with so much flexibility that we would likely be paralyzed without an overall understanding of what is happening.

This book focuses on the current technology and state of the art in communication networks and associated hardware and software products. It is my intention that it be a thorough comparative examination of the products, compatibility issues, and applications. This book should reveal how to best use network technology to implement network applications while reducing project backlogs and costs. It should show how to interface networks, PCs, and mainframe host computers.

Making the proper selections in network products and applications will ensure that networks are an asset, rather than a liability, for your organization. In order to accomplish this objective, you need to have an understanding of the state of the art of networks and the current products in the marketplace.

Different modes and definitions often pass for "compatibility," and it is important to recognize them while evaluating and selecting networks and add-on products. It is important to have a perspective on network hardware and software options and an understanding of the functions and differences in network-management systems.

This book will illustrate the interconnection options available for the popular micros and peripherals, such as modems. It will allow the prudent reader to know the differences among popular networks on the market and their features. It will increase your understanding of data communication options for PCs and mainframes and the networking and clustering options available for PCs.

Information-systems managers, analysts, design engineers, and technical staff whose responsibilities include the planning, acquisition, development, support, or use of communication networks and personal computers will better understand how to buffer the impact of networks and micros in their organizations.

Chapter 1 is an introduction to data communications. It is a review of the basic concepts, terminology, technology, and uses of various techniques for voice communications and switching. The operation of modems and multiplexers is also considered, as well as their characteristics and connection to networks.

Chapter 2 is involved with the principles of planning and designing networks. It surveys the key architectural principles used in today's leading networks and their application in different environments.

Network management is surveyed in chapter 3. It shows how you can relate vendor offerings to your needs and requirements and how you can define your network management needs more easily and accurately.

Network measurements can improve network efficiency with better ways of managing and monitoring. Chapter 4 surveys the newest management and monitoring systems and illustrates the best ways of diagnosing problems.

Network security management is surveyed in chapter 5. This chapter illustrates how a network can be secured with the latest in network integrity measures and how to evaluate and trade-off security issues.

OSI is becoming the backbone in standards and connectivity, and this new generation of standards is approaching rapidly. Chapter 6 centers on the specific standards that affect new products and system implementation.

New products often create new opportunities for communications. Chapter 7 presents a survey of product solutions for current and future needs in such areas as T-1, switches, network hardware, software, multiplexers, gateways, and bridges.

Emerging technologies represent a series of related developments and products that often lead to important breakthroughs. Chapter 8 is a current analysis of emerging technologies such as ISDN, Fax, and E-Mail.

1

An Introduction to
Data Communications

The rate at which digital devices such as personal computers, terminals, word processors, "smart" phones, facsimile machines, and quotation terminals are being used is increasing quickly. Exactly how all of these different technologies will interact with one another poses many problems.

Systems communications designers have had to accommodate these changes. Some vendors such as IBM and AT&T have found themselves in the building cabling business since telephone switches can control building lights and power. These directions were not foreseen in 1940 when the first data transmission took place, in 1964 when the IBM magnetic card typewriter ushered in a new era of office automation, or in 1975 when the first personal computer kit appeared in *Popular Electronics*.

The merging of computing and communications has been continuing at a furious pace. The two technologies used to be separate and as different as the symbols of the telephone pole and a punched card. First, machines were used to replace humans for calculating bomb trajectories, while other methods competed with the telegraph for long-distance communications.

In the post-World War II period, the invention of the transistor and its implementation in circuitry began a trend. Both technologies are now headed toward an all-digital, interconnected environment. Until recently, the merger of communications and computing was more easily talked about than realized. IBM did not really compete with AT&T in any core markets, and telephone system managers were separate from data-processing departments.

The technological changes were impeded by infrastructures that had to be repositioned, and users themselves were often given the task of how to manage the merged activities. The following are some trends that show how this merging has progressed. These trends also show that communications and computing are becoming the same entity.

- While 80 percent of personal computers in 1985 operated in a stand-alone mode, by 1988 70 percent were either connected to a local-area network (LAN) or part of a distributed processing environment in which PCs access host computers.
- Over half of the corporate telecommunications departments in the United States are now jointly managed with data-processing departments.
- Most telephone private-branch exchanges (PBXs) and key telephone systems being manufactured now are digital and have data-switching capability; by 1990 over half the installed base of PBXs and key systems will be digital.
- Most major vendors are now selling both telephone systems and computers. The same distribution channels are also supporting both types of products.
- Sophisticated office automation systems may offer the capability for voice annotation of documents. Many companies that began with word processing now also furnish voice mail systems, telephone switchboards, and networking.

As the technologies of communications and computing merge, corporations must merge their managing departments. Events like AT&T's divestiture and participation in the computer business and IBM's acquisition of Rolm have been noticed as signs of the times. Lap-top computers are used extensively as portable order-entry terminals for salesmen and electronic mail terminals for others on the road. Personal computers can be linked through a PBX to mainframes, and the PBX might manage both electronic mail and facsimile.

The bulk of corporate information passing takes place at the local level; for example, moving memos or notes to the desk next door or accessing data files. There are three separate networks in many offices: one for the telephones, one for computers, and one for environmental control. Each of these might support multiple subsystems. The least integrated and standardized of these network types is the one for computers. The communications alternatives are diverse and include:

- Large-scale local-area networking with up to hundreds of terminals, computers, and workstations.
- Application-specific local networking such as corporate video text.
- Smaller-scale local networking for sharing printers, disks, and data files between personal computers.
- Short-haul data communications using special switchboards.
- Local data switching using PBXs or key systems.
- Local data communications routed through a telephone company's central office.

- Mainframe control of data communications and access by dumb terminals.
- Departmental computing on multiuser systems with distributed applications and gateways to other networks.
- Application-specific networks with tightly coupled communications, such as those required by clustered word-processing systems or computer-aided design (CAD) systems.

One of the fastest growing sectors of office automation is the area of local-area networks. The total number of LANs installed worldwide in 1986 was 150,000, supporting about 2.5 million workstations. By 1992 the installed base could exceed 2 million and support over 20 million workstations.

An area growing even faster is for products that link multiple LANs together. Some examples follow:

- Local-area networks that might send their signals over already installed telephone wires rather than coaxial cables.
- Equipment interfaces built directly into products to make it cheaper and easier to connect them to local-area networks.
- The ISDN concept, which is designed to allow two telephone calls and one data transmission to run over the same wire that today handles a single call.
- Major computer and telephone system's now support common cabling schemes for buildings. Wires are run once in a building for both computer and telephone traffic. Initial cabling hardware and installation costs tend to be higher because the capacity must be present for future applications. Savings in rewiring costs to support change rapidly make up the difference.

Because of the way computer applications have grown, most corporations already have multiple network types in place. The applications themselves were justified, and often the most logical communications support was selected. Many users have discovered, however, that there are economies to be had in seeking to share some common facilities for separate networks.

From both the data-processing and office automation sides, the joining of technologies has had a profound impact. The networks themselves are becoming the processing entity, with computers acting as network servers and processing nodes. Information transfer and dissemination has become more important than information processing. *Connectivity*, the ability of devices to talk to each other, has become more important to users than the individual processing capabilities of the devices themselves. Data-processing managers are spending more time on networks than on classical data-processing problems.

Computer networks are required for communication between users involved with large file systems or distributed databases. These applications are common in banking, financial institutions, insurance, manufacturers, retail firms, distribution centers, and hotel and travel reservation systems. The data-processing requirements of all of these kinds of businesses require the transfer of data from one location to another. However, specific data-processing requirements often differ and include computer utilization, performance requirements, and input/output specifications.

Data collection might or might not be performed using a remote station to provide information to a centralized database. Message switching often is used for the processing and communication of messages between points or nodes via limited-channel-capacity links. The file management aspect can be accomplished using remote locations for updating a centralized file. Inquiry response systems can be used in some file-oriented applications. Here, a remote station might make inquiries of a centralized file, but it might not change that file.

Office automation applications might provide for shared tenant services. These can be based on word processing and telephone access at various sites.

ANALOG TELEPHONE SIGNAL CONCEPTS

Most of the early PBXs were designed for analog voice traffic. These analog signals appear to be continuous, as shown in FIG. 1-1, because the signals can take on all possible values between the lowest signal level and the highest. Binary or digital signals use only two possible values, representing the binary digits (or bits) 0 and 1 as shown in FIG. 1-2. The digital signals can be used to encode any desired number or letter. These letters or numbers can be represented by analog speech signals. A group of bits is often used to represent a

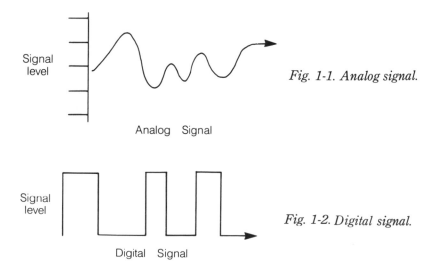

Signal level

Analog Signal

Fig. 1-1. Analog signal.

Signal level

Digital Signal

Fig. 1-2. Digital signal.

binary number or letter. The bits correspond to the value of the signal at a given instant.

The specific values represented by a group of binary numbers can be assigned according to any desired scale with either fixed steps between successive values (a linear scale) or variable steps (a nonlinear scale).

The analog telephone converts variations in sound pressure into variations in an analog electrical signal. The number of variations that occur sets the frequency of the signal. The speech that is produced can be considered as the summation of a number of pure tones that occur simultaneously. The difference between the maximum and minimum frequencies represents the bandwidth. The human voice contains signals of 100 Hz to 5000 Hz. The telephone system limits these frequencies to 300 Hz on the low end and to about 3500 Hz on the high end, producing a bandwidth of about 3200 Hz. This is enough to allow the transmission of recognizable speech. Analog speech can be represented in digital form following the steps of sampling and analog-to-digital conversion.

PRIVATE BRANCH EXCHANGE

Many remote stations have used terminals connected to facilities leased from the public telephone network. The public telephone circuits originally were designed for voice service and have limited data-carrying capabilities. The error rate may be greater than 10^4 times the error rate that is possible on direct digital links.

The cost of leased circuits can also be higher, but it can be reduced using multiplexing to service a number of terminals. A lower investment is also possible using smaller units and incorporating a structure tailored to the needs of smaller user groups.

The software applications are also available to serve a wider range of functions. More users can access the computer and this access requires that local workstations be close to their desks.

Asynchronous terminals have been used widely because they require less complex protocols than synchronous terminals. Software terminals are available through a number of sources and there is a wide range of equipment.

A common asynchronous terminal interface is based on the RS-232 standard. It is useful for supporting terminals located near the host. This same standard is also known as *V.24* by the International Telegraph and Telephone Consultive Committee (CCITT). It can support transmission at rates up to 19.2 kbps up to a distance of 50 feet. Remote terminals utilize devices identified as data circuit terminating equipment or DCSs.

ASYNCHRONOUS AND SYNCHRONOUS TRANSMISSION

Asynchronous data transmission is common for communication data rates below 19.2 kbps. It is used in most ASCII terminals.

The internal timing references, or clocks, of the transmitter and the receiver operate independently. The transmitter initiates the transmission by sending out a start bit, and then transmits the data at a fixed time interval.

The receiver detects the start bit and samples the data stream at fixed time intervals. The transmitter and receiver must be set for the same bit rate in order for the data to be interpreted correctly. The transmitter usually sends one or more stop bits to indicate the end of the data. The short bursts of data allow the receiver to resynchronize its clock at the beginning of each transmission. Long data streams need a high clock accuracy in order to remain synchronized over the longer data transmission.

Synchronous transmission is better suited for long uninterrupted bit streams at higher data rates. The clock information is encoded along with the data, which allows the receiver to adjust its clock rate. The Manchester-type of code, which provides two transmissions per bit, is often used to send long streams of data without resynchronization.

Synchronous transmission standards group the bits into frames that have a specific pattern to indicate the frame boundaries. The receiver scans the incoming data stream and uses this pattern for synchronization.

In the T1-D3 standard, for example, 24 PCM-encoded 64 kilobaud per second voice channels are multiplexed into a single 1.544 megabaud per second stream. Each frame holds an 8-bit sample from each of the 24 voice channels plus a framing bit for synchronization. This results in 193-bit frames that are repeated 8000 times each second for the 1.544 Mbps rate. Rates for synchronous transmission can range from 1.2 kbps up to the T4 rate of 274.176 Mbps. T1 transmission is often used between PBSs and hosts.

Synchronous transmission can be used for transmitting blocks of data in which the beginning of the data stream is used for synchronization. IEEE Standard 802 defines rates of 1, 4, 5, 10, and 20 Mbps for synchronous transmissions. The errors in these block transmissions can be checked using a block-check character (BCC) or cyclic redundancy code (CRC) at the end of the transmission.

VOICE COMMUNICATIONS

The U.S. public switched-telephone network has the capability to connect over 170 million users. This switched network has a uniformity and general ease of use that has made it a model for many data networks. The main reason the public telephone network appears easy to use is that the interfaces have been completely standardized. This includes the telephone equipment, the wiring, the numbering system, and the signals provided to users during calls, such as the dial, busy and ringing tones.

The internal operation of the network appears to be transparent to the user, and after the dialing sequence is complete the network takes over the responsibility for routing the call and maintaining the connection without any user assistance.

This standardization of voice systems has been due primarily to government regulation over the years and the industry domination of a single supplier, the former Bell System. Even with the recent deregulation of the industry, these standards will remain in place, and equipment that is designed accordingly will operate properly on the network.

Local communications products evolved in a freer environment, and they show more diversity. This is characterized by the many proprietary data communications standards that exist that are largely incompatible. Some of these can even vary from one product line to another. The differences include the binary codes used to represent information, the transmission formats, addressing and routing techniques, data rates, wiring methods, and connection set-up procedures.

The connections provide the paths for data exchange between the communicating parties. The connection appears to the user to be a physical path dedicated solely to their use. Not all connections will be supported by a distinct physical path because with multiplexing, a single pair of wires can support a number of simultaneous virtual connections that will all use the same physical connection. These virtual or logical connections are often used for data exchange among processors in a computer system. The software processor and the user cannot distinguish these virtual connections from hardwired links.

SWITCHING TECHNIQUES

Circuit switching and packet switching are two basic methods for allocating bandwidth. Circuit switching is commonly used to control and route voice traffic. A centralized hierarchical control mechanism is normally used that requires a global knowledge of the network.

Packet switching is more commonly used with data traffic. It can be used with distributed control mechanisms that require minimal information about the network configuration as the data packets move from source to destination. The same hardware often can be used to support packet switching for circuit-switching applications. Hybrid systems are also possible. In circuit-switching devices, a circuit represents a route or path between devices. Circuit switching provides the means to establish a connection between these devices. The data pattern remains fixed in time and space, and the path is fixed until the circuit is disconnected.

In order to establish a circuit-switched connection, it is necessary to locate an available data path, then assume control of it and allocate it to the exclusive use of the communicating links. This is sometimes called the *circuit setup overhead*.

Circuit switching tends to be more efficient when the amount of data exchanged is large compared to the amount of data that must be exchanged for the setup overhead. When the connection is made, there is normally no additional overhead required to maintain it, and the bandwidth and other resources are dedicated to the connection until it is broken. Circuit switching is efficient only if all of the allocated bandwidth is used.

Circuit switching is suitable for voice conversations that are relatively long (several minutes) compared to the setup overhead time, which is usually several hundred milliseconds. The ration between the data transmission time and the setup overhead should range from several hundred to several thousand.

Packet switching can be considered as a type of message switching in which the length of each transmission is limited to the maximum packet size. Longer messages are divided into a number of packets that are reunited into the original messages at the receiving end.

Packet switching makes use of the available bandwidth only when there is data to be transmitted. This is a more efficient use of bandwidth, but it requires more processing overhead for the transmission of each packet. The packets are formed by adding information to the beginning and end of each group of data. This is shown in FIG. 1-3.

A packet header identifies the source of the data and its destination. In some cases, it might identify the nature of the data or even provide billing and accounting information. The packet trailer holds the error-checking bits that resulted from the presence of incorrect symbols that can be present during the transmission.

Packet switching is more efficient in ''bursty'', transaction-oriented applications. Bursts of data are exchanged during the short period of time that the devices are logically connected during a session. A typical example of such an application is a distributed database. The communications costs for this type of application with packet switching are more likely to reflect the actual amount of data transmitted, instead of the total connection time. Packet switching is also more desirable in systems in which the users often switch rapidly between several applications or are required to use multiple applications simultaneously.

Delay considerations are an important factor in the relative efficiency of packet or circuit switching for a given application. Of interest is the delay between the time a device is ready to transmit data and the time the data is received at the destination. This total delay is made up of the transmission delay and the access delay. *Transmission delay* refers to the time required to send a message through the medium. This delay depends on the distance that

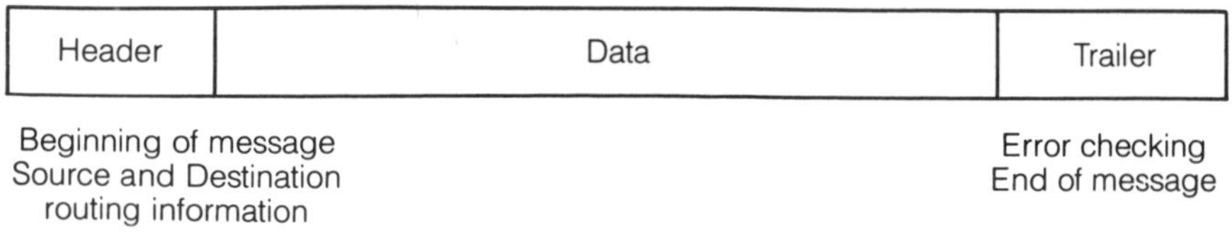

Header	Data	Trailer

Beginning of message
Source and Destination
   routing information

Error checking
End of message

Fig. 1-3. Typical data packet.

the message is sent and the bit rate for both packet and circuit switching. *Access delay* is the time that is required to access the transmission circuit before transmitting data. Circuit switching allows the circuits to be allocated to a connection immediately following the initial setup time. Packet switching can also suffer from an access delay prior to the transmission of each packet if the system is in use or if there are other users waiting with a higher access priority.

Some types of packet networks, such as those that use token-passing, have a deterministic maximum delay that can be incurred by a single user. This depends on the number of stations and the distances between them.

The use of protocols at the third and fourth layers in the OSI model can provide virtual connections that act as circuits for packet switching.

Networks that use the CSMA/CD access method have a maximum delay that is probabilistic. Long delays can be minimized by restricting the number of users. It is possible for the delay to increase to the point where no data gets through for a large number of users.

A type of fast circuit switching is available in circuit-switched systems. It uses an improved circuit setup procedure with a delay that is less than the typical 100 to 500 milliseconds.

Space-division switching is a technique that was used in many of the early manual and electromechanical PBXs. Space-division switching networks dedicated a unique physical path for each conversation. A full-duplex physical path was employed that used a continuous analog carrier for the duration of the call.

To illustrate this concept, suppose space switches are arranged to connect N inputs, and M outputs. If a single-space switching network is used to provide simultaneous connections between each input and any available output, a matrix of N × M switching elements is needed. Thus, to connect 300 users to any of 300 devices requires 90,000 bidirectional switching elements.

In practice, however, the probability of each user requiring service at the same time is low. Thus, voice PBXs provide for fewer paths than needed to connect each input to any other line or trunk. As the available switch paths become filled, the calls are blocked until a path becomes free. This is known as a *blocking system*.

The probability that a user will be blocked from service depends on the number of lines connected to the system and the average traffic through each connection. The total traffic is the product of the number of calls and the average duration of the number of calls times the average duration of each call. This can be expressed in seconds for a 1-hour period. Each 1 hour call is then equivalent to 3600 call-seconds or 36 ccs (hundred call-seconds).

For many business applications, a majority of the lines can require less than 5 to 6 ccs of traffic during the busiest hour. If the switching system has 3600 ccs capability, the system is nonblocking for 200 users because it can support 100 two-way connections of 36-ccs traffic.

A nonblocking system allows all possible combinations of lines and trunks simultaneously. These systems usually are not used in voice service.

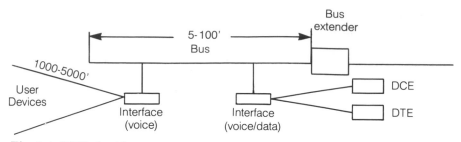

Fig. 1-4. PBX short bus.

The data bus in a PBX is a high-bandwidth data path that is shared among a number of devices. These can include voice stations, data terminating equipments (DTEs), or personal computer/workstations (PCs). The devices gain access to the bus through bus interfaces, which provide the device ports (FIG. 1-4). Digital PBXs often use these data buses to concentrate data before switching operations are performed.

The switching operations can be done in a switching network connected to the bus, or in the bus itself. These centralized buses are known as *short buses*. They are relatively inexpensive and may use parallel transmission to support very high bandwidths.

Communications between the user device and the bus can be through a link between the PBX port and the device. The data bus can operate at a high rate in order to support many users while the bandwidth of individual links to the devices can be much lower. The links must operate at the rates required to support a specified distribution of device types.

Long distributed buses with the full bandwidth are used as shown in FIG. 1-5. This type of bus is commonly used in baseband and broadband local-area networks. Long buses often use serial transmission over a single cable to reduce wiring costs.

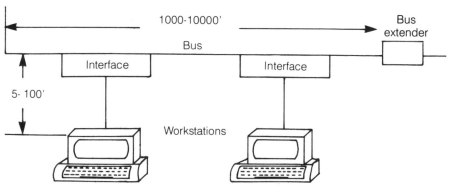

Fig. 1-5. Baseband or broadband long bus.

CONNECTIVITY CONSIDERATIONS

Most large organizations face the problem of interconnecting multiple networks. Two decades of growth in data communications has led to a bewildering array of networks. Businesses seeking a competitive advantage must link their nets to those of their customers and suppliers. Many network plans include net integration and consolidating into as few networks as possible.

A network utility is a single backbone that can carry all the organization's traffic and provide connectivity. It is necessary to reduce the number of networks and establish common protocols to make the elements of an organization's information infrastructure compatible. Differences between technologies are obstacles to total integration. Many protocol and architectural barriers will remain as the standardization process continues. The increased use of public and private networks, especially for overseas access, greatly increases the number of networks with which companies must interact.

The growth of distributed processing and the spread of low-cost minicomputers and microcomputers have increased the desire for full connectivity.

Cross-application data sharing and user demands for access to remote mainframe applications and databases from terminals and personal computers are forcing users to establish links between former application-specific networks.

Electronic mail, document distribution, and other services such as bulletin boards and on-line directories are additional reasons for giving users access to a shared network.

Along with addressing protocol standardization, adopting an internet architecture is one of the most useful plans for future network utility.

Today fewer obstacles exist to internetworking. Through internetworking, connectivity can be achieved without loss of control.

One approach to internetworking is one in which each internetwork link is treated as a special case that requires unique hardware or software. Access control and security functions are hobbled because of loosely defined requirements, and monitoring and control capabilities tend to be limited.

One problem is organization. There is a need to define who's responsible for the interfaces that span the separate organizations or departments. There is a need to develop an in-depth plan for an enterprise that comprises many different networks but creates the illusion of a single network to the user.

In many networks, gateways and bridges will become more important. Bridges act as simple relays that pass compatible protocols and formats between disjoined subnetworks.

Gateways act to translate between dissimilar architectures. In the network, gateways and bridges act as switches, making the routing decisions, buffering and revising data blocks, and implementing error and flow-control techniques.

An internet addressing standard is paramount to establishing large networks and can be the most important standardization goal for most large users. A single addressing format should be implemented that will allow any device to address any other device network. The International Standards Organization has been finalizing such a standard.

Data communication needs are changing quickly. This has been accelerated by the availability of networks that can connect computer-based systems to geographically dispersed users. The required data services are provided, in part, by modems operating over analog channels using a frequency-division multiplexing (FDM).

Rates of up to 9600 bps can be utilized over a single voice channel. Higher data rates require larger FDM bandwidths.

Digital transmission facilities can handle data requirements more efficiently; however, the early implementations that used PCM for digital telephony applications did not employ direct data interfaces, and modems were still needed to interface the data for PCM voice channels.

PCM channels can be used with modems at rates up to 9.6 kbps, but not all of the bandwidth is used. It is more efficient when the data channels are multiplexed directly with PCM channels.

A single 64-kbps PCM channel can be multiplexed to six channels of 9.6-kbps data. The submultiplexing of 64-kbps channels is often employed in digital transmission facilities.

The basic interface between data terminal equipment (DTE) and data-circuit-terminating equipment (DCE) in data transmission is shown in FIG. 1-6. The DTE is the user device, typically a computer, while the DCE is the communication device, usually a modem or multiplexer, that connects the user to the network. The interface is composed of three classes of signals: data, timing, and control.

The timing signal required will depend on the type of DTE used. Asynchronous terminals transmit data characters that are separated by start-stop pulses. These start and stop pulses indicates the beginning and end of each character, and no separate timing line is required between the DTE and DCE. Synchronous transmission transmits characters in continuous blocks without

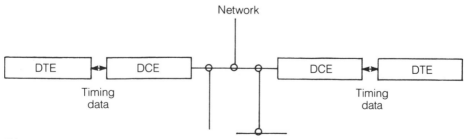

Fig. 1-6. Basic DTE/DCE configuration.

separate start or stop pulses. The transmitting and receiving terminals must operate at the same speed so the receiver can decode which bit is first in a block of data.

The timing lines in synchronous transmission can use one of the following three methods:

- Codirectional timing, which requires the DTE to supply a timing line to synchronize the DCE frequency source to the DTE.
- Contradirectional timing, in which the DTE is slaved to the timing of the DCE.
- Buffered timing, which uses buffers to allow each DTE to operate from its own timing source.

The buffers compensate for the differences in data rates between each DTE and the DCE. The buffered technique often is used when the line interfaces with several different terminals. These buffers can be a part of the DTE or DCE. They must be large enough to allow slip-free operation for a specific minimum period of time depending on the application needs.

COMPUTER NETWORK COMPONENTS

The components of a communications network that are in direct contact with the physical telecommunications transmission facilities are the modems, multiplexers, and concentrators. The circuitry, intelligence, and features that have evolved in these devices over the years have much to do with the basic operation and the techniques available for connecting networks to existing transmission facilities.

The rate in data symbols sent per unit time is called the *baud rate*. *Modem* was coined from the terms MOdulator and DEModulator. These components convert or modulate a signal into another form such as an analog tone for transmission and then demodulate the analog signal into a digital signal at the receiving end of the transmission line or link. These devices are true modems. The speed at which modems operate is given in bits per second. The data rate that is transmitted is given in bauds (data symbols/sec.) and can be equal to the bit rate in modems that operate at 1200 bps, or below, depending on the modulation method that is used.

There are other devices called *modem eliminators* and short-haul modems that do not actually perform the modulation and demodulation functions but are still called modems because they are utilized in the same way as true modems. These null modems and modem eliminators are essentially cable adaptors for connecting terminals or personal computers together using the RS-232C interface.

Amplitude modulation (AM) provided an easy way of achieving modulation and demodulation functions. The modulation function in large commercial transmitters is performed with vacuum-tube plate modulation. The plate of the output tube is controlled with the modulating signal and the carrier is injected at the grid. Now smaller transmitters use solid-state devices and employ the similar technique of collector modulation of the output transistor.

The detection of AM can be accomplished by passing the modulated waveform through a diode and then a filter to recover the audio signal. A simple AM receiver requires only a tunable antenna, diode, RC filter, amplifier, and transducer. The performance of such a simple receiver depends on signal conditions.

Commercial receivers employ superheterodyne techniques to improve sensitivity, selectivity, and fidelity. This involves shifting the received radio frequency signal down to an intermediate frequency and performing the recovery operations in this mode.

Frequency modulation (FM) allows improved selectivity and is less sensitive to noise and interference. These improvements are at the expense of increased bandwidth and circuit complexity.

The modulation of FM can be done with either direct or indirect methods. Indirect methods require integrating the modulating signal and then using this integrated signal to phase modulate the carrier. Direct methods use a carrier oscillator that is controlled directly by the modulating frequency. Monaural audio frequencies occupy the 0 to 15 kHz band, which are modulated and amplified.

FM stereo broadcasting employs multiplexing to provide the required channel separation using a single FM carrier. The so-called *matrix method* is utilized where the sum-and-difference frequencies are obtained with an analog summing and phase-shift network. The different frequency is then amplitude modulated at 38 kHz using a balanced modulator to produce upper and lower sidebands. A pilot signal is added to allow detection of stereo broadcasting. The detection of FM stereo can be accomplished by either matrix or TDM methods.

Pulse-amplitude modulation (PAM) is similar to a sampled-data system. The sampling switch can be either an analog switch or analog multiplexer.

The analog inputs are switched into a common transmission line using a time-sequence generator. One of the switch channels can be used to transmit the clock signal for synchronization during demultiplexing. The demultiplexing circuit is functionally equivalent to the multiplexing circuit, and all operations are performed in reverse.

Pulse-duration modulation (PDM) can be obtained from PAM with the addition of a voltage-controlled oscillator or multivibrator (VCO or VCM) to convert the amplitude of the pulse train to percent modulation of the fixed amplitude pulses. Another technique for PAM to PDM is to combine the PAM pulse train with a triangular wave or sweep and then clip this signal with a level detector to provide PDM.

Pulse-position modulation (PPM) can be obtained from PDM by performing a differentiation followed by clipping. PPM conserves signal power but it requires more bandwidth.

PCM is used extensively in modern data communications systems. Binary coded data is converted to frequency-shift-keyed (FSK) signals for transmission over a telephone link by a modem. Modems are then used to convert the transmitted PCM-FSK signals to binary data. Multiplexing is achieved by FDM of the transmission signals or by TDM of the binary signals before transmission.

Prior to the 1968 Carterphone decision by the FCC, hardwired connections to a common telephone link were not allowed. Data communications through the telephone network had to be accomplished with devices such as acoustic couplers that interfaced via audible tones with the telephone headset. Since 1968, independent manufacturers of data communications equipment have been allowed to pursue this previously closed market, resulting in rapid growth and increased capabilities.

Digital signals are transmitted by the selection of one of a number of wave forms that are used to represent the binary information. The transmitter function is to convert a string of data symbols (the binary digits or bits) into a sequence of physical symbols representing the data. The major receiver function is to make the correct decisions on the sequence of symbols in order to allow the full recovery of the data. The communications requirements of the channel through which the data is transmitted sets the nature of the methods used to represent the data.

Voice-grade telephone channels were originally designed to transmit analog voice signals in the frequency band of 300 to 3000 Hz. The signals used in modems must fit into this band and operate within the various limitations of the communications channel.

A major part of the cost in providing long-distance interconnections is the cost of transmission. Private leased facilities from a common carrier can be used. Public network channels can also be rented on a time-shared basis as in the direct-distance dialing (DDD) telephone network.

As discussed earlier, data communications tend to be "bursty" and non-symmetric in practice. There can be many quiet periods when little or no data is sent. The volume of data sent in one direction can be much larger than that sent in the other direction.

Many devices generate data that is only a small fraction of the total capacity of a voice-grade telephone channel. The cost of the channel per unit time is independent of the data rate up to the channel capacity so it is efficient to employ techniques that allow several low-speed devices to share a channel and to attempt to smooth the bursty and statistical fluctuations common in data communications. The transmitted data rate will then represent more closely an average rather than a number of peaks and valleys.

Multiplexing can be used to meet these objectives. For example, multiplexing allows the use of a modem with a capacity of 4800 bits per second for the transmission of sixteen 300-bps terminals.

A cost saving can usually be achieved with most of the available tariffs and equipment. Suppose the terminals are active only 5 percent of the time; a 4800 bps statistical multiplexer could be used to handle up to 320 terminals.

An intelligent multiplexer implementation will usually have a lower cost than the lines and modems that would be required to support 320 terminals. Intelligent multiplexers usually employ programmable microcomputers with their own memory. These can be programmed for functions such as data compression, automatic speed changing, protocol conversion, system diagnostics, and error control.

A *concentrator* is a device that not only performs multiplexing to one output port, but also performs switching and routing to output ports as well as data compression, code conversion, error control, and protocol functions. A concentrator is used in the same way as a front-end processor in that it relieves the main or master processor of some of the more routine communications tasks. In modern times, intelligent statistical multiplexers have taken over most of the tasks previously performed by concentrators. Concentrators now function as the switching node in large networks and provide multiplexing, switching, network management, and link protocol functions.

MODEM CHARACTERISTICS

During the 1970s, most modems were used for voice at speeds of less than 300 bps. Most of these were manufactured and installed by the Bell System, which is now broken up into ATTIS and the Bell Telephone operating companies.

There are several million modems now in operation, supplied by many different independent manufacturers. Speeds of more than 16,000 bps are available, and built-in features include handshaking, bandsharing, multiplexing, data compression, unattended autodial functions, and diagnostics.

True modems are characterized as either asynchronous with no clocking or synchronous, in which the data stream is synchronized to a clock. Asynchronous modems typically operate at speeds of less than 1200 bps, sending one character at a time. Modems that operate above 1200 bps are usually synchronous. Half duplex (HDX) characterizes a modem that operates in both directions but only one direction at a time. These modems must be switched between the transmit and the receive mode.

Full duplex (FDX) modems operate in both directions at the same time. Limited distance, or short-haul, modems can operate on wire-pairs for distances of up to 20 miles.

Many early modems were acoustically coupled into the telephone handset using a transducer. Some newer devices are not true modems but line drivers that operate at baseband, without modulation directly into the telephone lines. The following speed classifications are typical for modern modems:

- Low speed—less than 600 bps
- Medium speed—1200 to 4800 bps
- High speed—greater than 9600 bps
- Wideband—bandwidth over 16,900 bps

Modems also can be grouped according to their mode of transmission: point-to-point or multidrop, half duplex or full duplex, two-wire or four-wire. You also can classify modems based on the types of lines the modems will operate on: dial-up lines, leased telephone lines, or twisted pairs (which can be leased or owned). Two-wire lines are usually used in connecting to the dial-up public system.

The transmit and receive signals are sent together on a single pair of wires. They usually are isolated by a transformer using a balanced bridge circuit.

Most long-haul sections of the telephone network use a four-wire configuration. This allows two independent channels for the transmitting and receiving functions (FIG. 1-7). The four-wire circuit is subject to echo effects. Voice-grade channels generally employ echo suppressors when the echo delay exceeds 150 to 200 milliseconds. Echo suppressors introduce an additional attenuation in the direction that is not currently being used. If the modem is operating in full duplex, echo suppressors cannot be used. Most modems provide the required disabling tone, which is a single 400-millisecond frequency of 2010 to 2240 Hz.

Multidrop modems use a master modem that polls a number of remote modems in some defined sequence. The polling time will be a function of the response time of each modem to send when polled. This response time is typically less than 10 ms and represents the time interval between when the request to send (RTS) is received from the data terminal equipment (DTE) and when the clear to send (CTS), to which the modem responds, is sent back to the DTE. This is shown in FIG. 1-8. Polling can also be used with the dial-up network, but the total dialing and connect time exceed 20 seconds on leased four-wire lines.

Modems can be physically independent modular units or plug-in printed circuit boards (PCBs). Large-scale integrated (LSI) chip sets are used in many units that are employed in modem racks, terminals, multiplexers, concentrators, and personal computers.

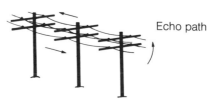

Echo path

Fig. 1-7. Four-wire transmission.

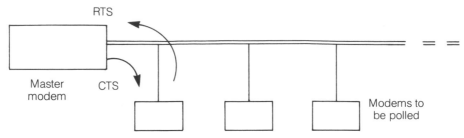

Fig. 1-8. Multidrop modem polling.

Stand-alone modems are usually self-contained with their own power supply and function and error lights. Automatic call units (ACUs) are often employed to store lists of numbers that can then be dialed automatically.

Switching the line between data mode and voice mode is available, and a standard interface is normally used to pass data, control, and autodial instructions from a terminal or personal computer. Operation at more than one baud rate is common, and compatability with other modems currently in use, such as the Bell 212, is important.

Many newer smaller modems have automatic dialing. Automatic dialing functions in earlier modems used on the DDD network required the use of a separate ACU. These were stand-alone units with their own interface to the modem and to the computer.

Inverse multiplexing and multiport operation are other newer features. Multiport or split-stream operation is a type of multiplexing that is done at the input to the modem. It permits several different data streams that can be at the same or different rates to be combined into one signal. V.29 modems, which operate at 9600 bps have this feature.

Autodial can be used with polling, or it can be used as a dial-up/backup feature when a leased line opens. Some autodial modems use a list of numbers that are programmed by the user through an RS-232C interface.

The dialing can be keyboard controlled or controlled by software. The directory can also be stored in non-volatile EPROM so it is retained during a power interruption. A linking feature can be employed that replaces the first number dialed with the next number if the first call is unanswered or busy. This provides a basic technique of least-cost routing.

One application of multiporting is combining four 2400-bps independent data streams into one 9600-bps output. Another example is the combination of two 2400-bps into one 4800-bps stream.

Inverse multiplexing can be used with several voice-band modems with several lines for transmitting a single data stream at a higher rate than can be accomplished with a single line. A 19.2 kbps data stream can be split using inverse multiplexing by sending the odd bits occuring at 9.6 kbps through a 9.6 kbps modem and a voice-grade line, while the even bits are sent through

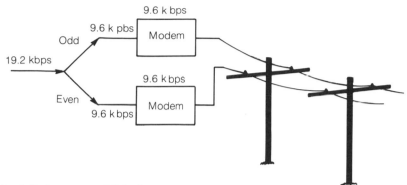

Fig. 1-9. Inverse multiplexing.

another 9.6 kbps modem and a second voice-grade line as shown in FIG. 1-9. When the input data rate is less than 19.2 kbps, and one or both of the modems is multiport, the excess may be sent through the other modem ports.

A major reason for using inverse multiplexing is the gap that appears in the available channels above 14.4 kbps. Another is fail-safe backup because if one line fails, the other can continue at a reduced rate. Problems can occur in the reconstruction of the bit stream from two or more lines because of different and changing propagation delays. This problem can be overcome by using some form of elastic buffering or dynamic alignment of the bit streams. Transmission over hundreds or thousands of miles of terrestrial wire or by satellite facilities can produce delays of up to a second.

Many HDX modems have a low-data-rate reverse channel that is used for signalling and/or automatic repeat-request (ARQ) acknowledgments in link protocol systems. Modems can also be used to alternate data and voice on a single line. Synchronous-to-asynchronous conversion can also be included in some modems.

Diagnostic techniques include monitoring local and remote modems, as well as the line, using a remote digital loopback. This often is used to test the modem-line combination and for fault isolation of the lines. Other techniques include internal eye-pattern generation to check the amount of phase jitter and noise. Essentially, all modems have built-in and standardized handshaking protocols that are used to establish the initial synchronization.

Many modems are compatible with those manufactured by AT&T. These are known as the AT&T or Bell type of modem. Many of these are listed in TABLE 1-1.

Many manufacturers provide modems that are compatible with the Bell 103, 113, and 212 (see Appendix). Other standards in use include the CCITT V.21 recommendation for 200 bps FDX modems in the switched telephone network.

The United Nations Consultative Committee on International Telegraphy and Telephony (CCITT) publishes international standards. V.24 is virtually the

Table 1-1. AT&T or Bell-Type Modems.

Type	Date Rate (BPS)	Transmission Mode	Clocking	Modulation Method
102 A,E,F,J	0 to 300	HDX or FDX	Asyn.	FSK
113 A,B,C	0 to 300	HDX or FDX	Asyn.	FSK
201 A,B,C	2400	HDX	Syn.	QPSK
202 B	1200 DDD 1800 Leased	HDX	Asyn.	FSK
208 A,B	4800	HDX	Syn.	PSK
209 A	9600	HDX	Syn.	QAM
212A	300 or 1200	FDX	Asyn.	QPSK

same as RS-232C. It is similar to the Bell 103 technique. Another standard is the CCITT V.23 recommendation for 600- to 1200-bps HDX modems. At the higher rates of 4800 and 9600 bits per second, many modems are also compatible with the CCITT international standards designated as V.27, (U.S. Federal Standard 1006) and V.29 (proposed Federal Standard 1007). CCITT V.22 is another standard for 2400 bps on dial-up lines.

The maximum theoretical transmission rate on voice-grade lines can be as much as 30 kbps, but the techniques required are expensive and include the use of special modulation, automatic equalization, and phase-tracking.*

PROTOCOLS

In a typical network application, there may be several different kinds of data sources. These may include different mainframe, minicomputers, or PCs. These source devices might need to exchange data with several different types of receivers including text or graphics terminals and PCs. For these three output devices (unless a standard shared protocol is used) nine different protocols will be required for all the possible connections. Each of these protocols will serve a special purpose.

A more reasonable solution for interconnecting the equipment utilizes a standard protocol that would require only one connection for the six implementations made up of the three sources and three receivers (FIG. 1-10).

Multilevel protocols use several single levels of protocol as layers of protocol in a hierarchy. These multilevel protocols allow the separation of functions that are required in most complex systems. In these systems, the responsibility for resource management is segregated into the various areas required by the different resources. A multilevel protocol can help to support modular changes in the network because each level of the protocol tend to act as a separate module.

---

*Modems that can evaluate such parameters as the quality of the line and adjust speeds accordingly have been operated at speeds greater than 21,000 bps.

Fig. 1-10. A standard protocol implementation.

Multilevel protocols are more flexible and allow special-purpose protocols to be substituted or used in parallel with standard protocols as needed. The multiple levels can still be general in nature while maintaining the advantages of single-level protocols for special applications.

A one-level protocol has the advantage of being simple to implement. Most one-level protocols are application-dependent, special-purpose protocols developed for a specific application.

Other questions are whether the protocols should be transparent to the user or whether they should perform a virtual communications function. A transparent protocol is not noticed by the user, yet it exists, providing a function that the user does not need to control. The main advantage of a transparent protocol is that it makes user operations simpler.

A virtual protocol performs a tangible service, such as creating a virtual file, but the virtual protocol exists only as a logical entity, and it has no physical existence.

When two protocols of different levels communicate, the lower-level protocol first must accept the data and control information from the higher-level protocol before performing any operations on it. The lower-level protocol typically treats all of the data and control information as data and adds its own set of header and trailer controls. As the data moves through the protocol layers, it will collect similar control-data envelopes.

A host-to-host protocol will accept application data from the user program and add its own control information. As the data travels further, the link control procedure will add more information. This hierarchy of protocols is similar to the nesting of levels in a complex operating system.

Protocols can be used for a variety of functions. In many networks, more than one protocol can be employed with each protocol supplying a specialized set of functions.

There are some basic differences between protocols and interfaces. The protocol is the set of rules for communications between the similar processes, while the interface applies to the set of rules (including connections) between dissimilar processes. An interface is the physical connection between two devices or processes, while a protocol is a logical concept. Consider a system with an interface between a host computer and a switching node. In this sys-

User
Logical communications
Interface
Host
Interface
Physical connection
Node

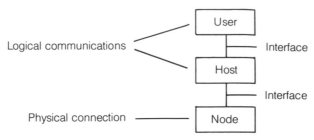

Fig. 1-11. Protocols and interfaces.

tem, there may be two protocols: a host-to-node protocol and a host-to-host protocol as shown in FIG. 1-11.

MODEMS-TO-NETWORK CONNECTIONS

The physical connections for modems and DTEs are examples of standard interfaces. A 25-pin RS-232C/V.24 interface is typical. This interface is the lowest or physical-layer protocol in the International Standards Organization (ISO) model for the Open Systems Interconnect (OSI) reference standard for computer networks. The RS-232C standard is described in the Appendix.

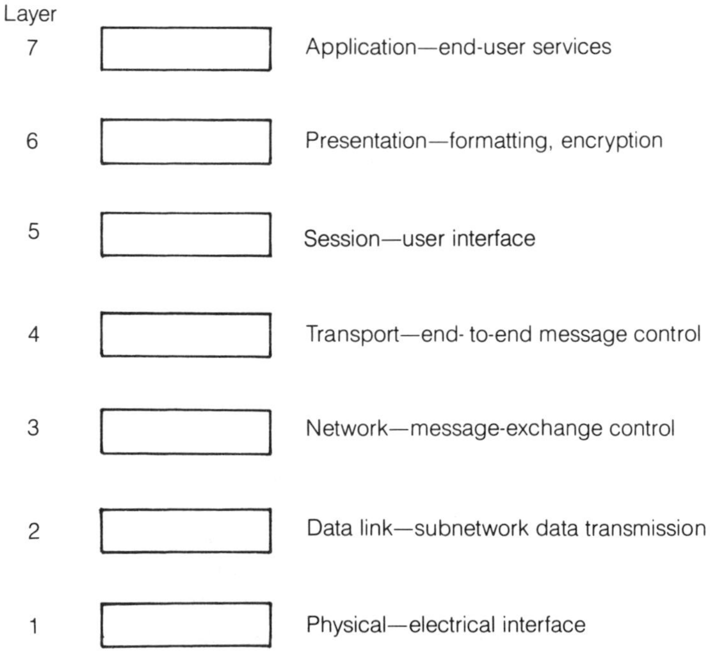

Layer
7
Application—end-user services
6
Presentation—formatting, encryption
5
Session—user interface
4
Transport—end- to-end message control
3
Network—message-exchange control
2
Data link—subnetwork data transmission
1
Physical—electrical interface

Fig. 1-12. In the ISO model for Open Systems Interconnection (OSI), data passes physically through layer 1; all other layers transfer information through virtual connections.

The physical interface protocol of the ISO model specifies the pin connections, functions, and electrical characteristics, including voltage and current levels and recommended cable types and lengths. FIGURE 1-12 illustrates the basic ISO model, and the Appendix illustrates the difference between the EIA-RS-232C and CCITT-V.24 designations.

Some modems operate at frequencies above the voice band in an FDX mode on telephone lines. These are local-area networks that are generally unique to a particular building or facility.

In these networks the lines are unloaded, and band-limiting filters are not installed by the telephone company except where the lines are connected to the public network. Some of these modems operate with FSK, 9600-bps data at about 40 MHz in one direction and above 75 MHz in the other direction. This form of modem can also be used in an analog PBX to provide a low-cost voice/data local-area network.

When modem polling is unsuitable, modems can be connected in networks. Both standard and statistical time-division multiplexing can be used in master-slave configurations.

MULTIPLEXERS

The basic reason for using multiplexing is to save line and modem costs through the sharing of a common channel. Multiplexing is typically used by the common carriers in their networks for voice transmission. This is a form of trunk muliplexing.

MULTIPLEXING METHODS

Multiplexing involves using technology that allows multichannel capability from a single resource. Multiplexing can be accomplished in either the frequency domain, as in frequency-division multiplexing (FDM), or the time domain, as in time-division multiplexing (TDM).

The origins of multiplexing go back to early radio and telephone communications in which a transmission medium, such as a broadcast band, was subdivided into channels, with each channel occupying a distinct frequency.

FDM

Frequency-division multiplexing involves modulation techniques that multiply the message function by a higher frequency (usually a sinusoid), referred to as the *carrier*. The sinusoidal carrier has a unique and discrete spectrum that allows noninterfering broadcasts of many communications channels or stations. The selection of sinuosoidal carriers can also be accomplished with simple tuned circuits.

NETWORK CHARACTERISTICS

A typical network will be made up of subnetworks as required for switching modes, transmission facilities, and the subscriber equipment, including hosts, terminals, and processors. The transmission equipment can include the multiplexers and switches.

In order for the data to travel over the network, there can be specified levels of performance that are necessary to maintain network applications and functions. This can include resource sharing among hosts, terminals, and other equipment. The protocol design is often the key to effective resource sharing among the network devices. This is because most processes within the host computer communicate with other host processes or with processes in terminal handlers using a specified protocol. In a similar way, the network transports data using the low-level protocols.

Virtual protocols are often used in complex systems in which the higher-level protocols are built on top of the lower-level protocols. The virtual communications are established by the lower protocols with an end-to-end protocol used to control the overall integrity of transmission.

The link control procedures as covered in level 2 of the International Standards Organization (ISO) model can be considered the core protocol for most networks because it is the most basic. The important link control procedures include sequencing, flow control, data transparency, connection and disconnection, error control, and failure recovery. The most popular link control protocols are HDLC, SDLC, and BSC (BiSync). HDLC is the most popular in new designs of these three.

A basic feature in link control is the provision for data transparency. Both data and control messages can travel through the link, so it is necessary to differentiate the data bits from the control bits.

The link control protocol must be able to distinguish when and where a data block begins and ends without restricting the data content. The protocols will assign a control meaning to specific data patterns. The beginning and end of a block of data will be identified by some specific pattern of ones and zeros.

Two basic methods for achieving data transparency are *bit stuffing* and *byte stuffing*. Bit stuffing is used in HDLC and byte stuffing in BySync. Byte stuffing operates with a pair of characters and bit stuffing with a series of 8 bits. The transmitting hardware adds the bytes or bits, and the receiving hardware removes them, thus preserving the transparency of the original pattern of data.

One technique doubles the required control byte, so when the receiving hardware recognizes two control bytes in a row, one of them is automatically removed. Both bit stuffing and byte stuffing work in a similar way.

Connection and disconnection also are done as part of the link control procedure. As a connection is established, the devices involved in exchanging information can also exchange various types of identification in a handshaking phase. This is also the time when various parameters important to the data exchange will be set, including provisions for disconnection and reconnection if and when errors should occur.

The link control technique used for recovering from failures usually involves some form of periodic test of the link operation. If the link should become inoperative and fail the test, both ends of the link must contain recovery procedures. When both ends have been notified that the link is malfunctioning, they start to establish a new virtual connection. The mechanism used must be fail-safe so that the link can continue functioning even in the presence of errors in transmission of the data or control information.

Error control is a feature which allows the link control to perform failure recovery and maintain accurate communications under normal operating conditions. The link control procedure provides this through the management of transmission errors.

A variety of methods are available for error control. One approach is to use an error-detection mechanism for each transmitted message and require retransmission by the source if the receiver does not acknowledge an error-free reception.

The retransmission can be initiated by either a negative acknowledgment from the destination, if the message is not received, or by a positive acknowledgment if the message arrives.

A timer in the source can also be used to retransmit messages when positive acknowledgments are not received within a set period of time. Positive acknowledgments are more reliable because the negative acknowledgments themselves can be lost.

A fundamental method of detecting errors is *parity checking*. One bit of the data is reserved to indicate the parity, which is based on the number of ones or zeros in the data string. Other more complex techniques for detecting errors include computing bit summations such as the *cyclic redundancy checksum* (CRC).

The source can identify each of the blocks it is transmitting by a control number. This number can be set to count from zero to some set limit and then recycle. The destination can use this sequencing to detect missing blocks or duplicate blocks. Messages with numbers lower than the lowest expected number, or higher than the highest, become duplicates.

If a block does contain errors, the destination device might ignore the block. Then, after a predetermined amount of time, (usually the time required to receive a positive acknowledgment), the source can retransmit the block.

Traffic sequencing is not a basic function of link control, but it can be used to ensure that the traffic flow leaving a particular link occurs in sequence with the traffic entering it.

A large switching network can have many links, and sequencing is usually more efficient if it is done at the final destination. Large centralized networks often use sequencing at the link control level. The blocks are assigned a unique sequence number, and this number is used in detecting missing and duplicate blocks as well as maintaining status information.

Older systems that had only one or a few links work well with the early link control procedures, which were designed for only one link between the host computer and the peripherals. As these systems grow in complexity, they soon outgrow the original link control sequencing.

Flow control here involves the sender's transmission rate and the receiver's capability to accept traffic using the link control procedures. One fundamental technique uses the explicit allocation of resources. Here the receiver explicitly notifies the sender that it can accept traffic, and a designated storage area is usually reserved for messages to be transmitted. A simple flow control that is used in some systems consists of a basic on/off control scheme with its inherent limited efficiency. Many actual flow-control techniques use some type of feedback.

In one type of feedback, the sender transmits messages until the receiver begins to reduce the amount of allocated space.

In some systems, the receiver notifies the sender by specifying some transmission rate that can be supported. This prevents the receiver from becoming overloaded with traffic. The flow control mechanism then allocates the number of messages to the sender based on the receiver's instantaneous reception rate. This requires a knowledge of the window in which a limited number of messages will travel from source to destination to approximate the flow control.

Another type of flow control within the link control system requires discarding messages when they cannot be accepted. This method relies on a higher level of protocol to retransmit the messages at a later time.

Flow control can also be performed through error-control mechanisms. One simple technique is not sending positive acknowledgments. This provides flexibility that is not possible with simple on/off control.

2

Planning and Designing Communication Networks

Once it has been decided that an integrated network might be the solution to a heterogeneous mix of equipment that has been acquired over the span of several years, the first step in evaluating the feasibility of such a project must be evaluated. The first step in this process is to determine how this area network fits into the organizational structure.

EVALUATING NETWORK NEEDS

A clear vision of the present unconnected informational network is needed. The major considerations are shown in TABLE 2-1.

Network technology is changing rapidly, and many new products are always appearing. As a result, complex decisions must be reached before a commitment is made.

The methodology used should result in the selection of a network that will not hamper growth. It should satisfy almost (but not always all) all of the desired requirements within current and projected budget allocations.

Many different networking products are available, and choosing the best solution can be a difficult task. Without recognizing and carefully quantifying network requirements, proper equipment selection will be difficult and often confusing.

Rather than initially comparing available products, planners must first assess the present needs and future requirements of the organization. When this step is complete, the next step is one of familiarization/education of network product technologies. Next can be an initial in-house selection/recommendation of those product technologies that appear to be most suitable. An evaluation or pilot program of specific products can be undertaken before committing to a final plan selection.

Comprehensive network planning requires a thorough assessment of specific needs. Network users can have special needs, such as specific graphics

Table 2-1. Network Characteristics.

The computers, terminals, and peripherals to be connected

The data rates and protocols to be used

The information that would be generated, stored, and distributed to network workstations and other intelligent devices

The information that would be protected from general access and require controls

The databases locations and how they can be interconnected

The equipment projected to be added to the network later*

---

*Emerging technologies that integrate video and voice with data transmission should be considered as part of the last item.

---

requirements, high-volume printing, or printing on special forms. Such techniques in area networks as the use of common printers, each loaded with a different form, should be considered as an effective strategy for minimizing hardware and maximizing efficiency in the network. The sharing of hardware peripherals such as printers requires a consideration of the duty cycles involved. A device with a recommended duty cycle of 55 percent should not be used 85 percent of the time, or there will be an excessive amount of downtime and repairs.

The following steps are typical for an area network implementation:

1. Perform an initial needs assessment to define the implementation plan.
2. Investigate the available network technologies and generic products and compare costs.
3. Request vendor information and research recent product introduction.
4. Perform a preliminary comparison of the viable generic solutions.
5. Prioritize future requirements that require solutions.
6. Develop a request for proposal based on the highest priority requirement.
7. Evaluate vendor responses and develop a ranking matrix with costs, comparing the design criteria with the associated rank values.
8. Request final bids from the vendors with the highest ranking.
9. Select the final network vendor(s) and negotiate contracts.
10. Plan pilot installation, acceptance test, training, and support.
11. Begin full installation after successful pilot installation.
12. Perform system tuning/optimization based upon the initial traffic statistics.
13. Plan for future growth and document the project accomplishments.

Area networks can be grouped into three basic types:

1. Terminal-to-terminal networks, which can include Telex and facsimile devices.

2. Terminal-to-computer networks, which use protocols such as SDLC and 3270. These are typically low-speed, star-configured networks.
3. Computer-to-computer networks that can connect several hundreds of computers at several hundreds or thousand of kilobits per second over distances of several hundred meters.

In many cases, these networks will be networking personal computers. Shared file systems on large Winchester disks allow group collaboration, and the Winchester disks provide a much faster response time and can be more cost-effective than individual floppy-disk systems. Centralizing the disks also allows better maintenance away from the users and more reliable backup procedures.

Area computer networks allow personal computers to share printers, thus reducing the cost-per-user of expensive high-quality printers in a facility. Networks allow the shared use of laser printers for in-house publication. Other peripherals that can be shared are modems and gateways to other networks. The need for multiple printers and modems is thus reduced and can, in part, justify the cost of the local network.

Area networks allow cost-effective electronic mail between personal computers. This can be built into the network file system, or can occur through a gateway into a central electronic-mail system.

Network technologies can be grouped into three basic types: PBX, cable TV broadband, and the baseband types such as EtherNet. PBX technologies usually provide only 4.4 kHz of bandwidth. Digital transmission is generally supported with RS-232C or V24 terminal communications. PBX is most suitable for terminal-to-computer networking.

Broadband networks based on cable television technology can provide up to 400 MHz of bandwidth. The bandwidth is usually frequency-divided into a number of channels, which makes it suitable for bundling together lower speed lines. The faster broadband networks utilize channels of up to 128 kbps.

Digital baseband networks generally operate at about 10 megabaud. They can provide their entire bandwidth, if required, between any two points. These types of networks are most useful for handling bursts of data that are generated by communications between computers.

Different network technologies can coexist in a system with each serving a different set of needs. Interfaces are available between the various technologies, such as gateways to EtherNet. The International Standards Organization's seven-level network model* provides a way of defining the compatibility of the layers of network connection and protocol. Protocols such as IBM's SNA, DEC's DECnet, and Xerox's XNS are higher-level architectures for levels 3

---

*The ISO model sets standards for the exchange of information among systems that are open to one another due to their mutual use of the standards. The standards provide a formal, logical structure for the interactions and functions required to provide communication services to the user, as discussed earlier.

through 7. An SNA network can include EtherNet as a low-level transmission medium. A variety of local-area networks at levels 1 and 2 such as EtherNet, RS-232C, 3270, or IBM's token-passing network, could exist within SNA.

The higher-level protocols are where most problems occur. However, even when one vendor's EtherNet shares a cable with another vendors's EtherNet, inconsistencies can still occur. Although the microcomputers might be able to communicate with each other, they might not be able to use such features as EtherNet file or printer servers from other vendors. The problem of compatibility must be solved before PCs can exchange electronic mail and utilize peripherals on these mixed systems.

The bottom-up approach for networking personal computers is based on the view that since no one network technology is able to accommodate all of the needs in an organization, it can be effective to implement local-area networks as the need arises from the user's viewpoint, starting from the bottom of the organization structure.

The higher-level protocol standards force users to select personal computers and applications before selecting a network. The applications and the PCs then determine the higher-level protocols and set the choice of local network. This application-driven approach is controversial because it implies a bottom-up decision making process. The bottom-up approach addresses specific groups of users with their unique needs, rather than wiring an entire building. In a system of multiple networks, compatibility can be achieved using gateways between the networks.

Communications managers seeking management support for major changes in their communications operations should demonstrate to management how the new scheme addresses a particular business problem or opportunity by preparing an internal business proposal. Communications managers have lagged behind their MIS counterparts in presenting proposals that relate their departments' strategies.

Communications managers often view themselves as intermediaries between their organizations and the telephone company, rather than as active players in the corporate business strategy.

They should relate their ideas to the company's overall business plan. The company's business may vary. Communications is a tool for company performance. Before working on a request for proposal to be sent to vendors, managers must prepare an internal business proposal for management.

For many communications networks, a zero-phase proposal is an effective initial technique of presenting plans. This is a simplified business proposal designed to test the waters to major changes in an organization's communications operations. It states the idea and its benefits.

Although the zero-phase proposal is essentially a trial balloon aimed at testing initial interest, it can be incorporated into a more inclusive business proposal. Most of the principles of proposal presentation remain the same, whether the proposal is long and inclusive or simply a zero-phase proposal.

A common fault in trying to sell equipment purchases and systems changes to management is an overemphasis on cost savings. Cost saving might be a relatively low priority. It might be possible to save a great many dollars per year at a large corporation, but that is not really all that important if what is really needed is to increase sales.

Promising to improve the quality of the organization's communications service will not generate any interest if the gains interest stemming from the improved service are not clear. In many companies, the sales and marketing divisions have information on the average dollar amount of business generated by each incoming and outgoing telephone call. Information of this type can be used to demonstrate the benefit of improving communications service.

In any business proposal, even the most rudimentary introduction should include the necessary justification and cost-evaluation. This section of the proposal, should expand on the reasons why the company should invest in a project, while relating the justification to business objectives. In the zero-phase proposal, the figures can be believable but "soft", coming as they do from experience and reports such as station-message detail-recording reports. More inclusive proposals require more accurate financial information.

All proposals should include an executive summary. This is generally the most widely read section of the report and can attract the interest of members of the organizations who at first do not realize the importance of the project.

When it comes to developing strategic communications plans for competitive gain, there are several ways for creating the plan. In some cases, upper management dictates strategies to departmental managers. Lower-level managers may try to "sell" their concepts to the executive team. In other cases, teams made up of all management levels and corporate departments might work out the details together.

The ways in which companies develop strategic plans are as diverse as the many methods of management. Members of the following three groups should possess some vision in regard to strategic information systems: top management, line management, and information management. Top management sets the firm's strategic direction. Line management functions are in close contact with the firm's strategic targets, and information management must access the strategic significance of new technologies.

Establishing a staff unit that is isolated from direct line functions has merit for mature organizations. It allows activities such as research or planning to proceed without the demands of the constant brushfires a line manager faces. It focuses on the strategic issues and allows a clearer review of options.

Many organizations allow former line managers to head up these staff positions. Someone who has survived the experience of implementing, managing, and maintaining communications systems brings several advantages to the staff function.

They know from former responsibilities what the crucial needs of research and development are as well as the concerns of staff and actual users' needs.

This type of manager will not be as easily misled by the promises of vendors proposing magical solutions to complex problems. A practical-minded realist is needed who is not likely to propose improbable procedures, or unworkable or abstract solutions.

The creation of strategic applications might begin at the departmental level. This mode involves translating technical applications into business objectives, which are then included in the division's overall planning.

Department strategies, objectives, and capital plans are incorporated into the overall corporate plan. The relationship between senior management and the department is crucial. It should not be an "us versus them" relationship. It should involve discussing the idea and exploring the possibilities. The presentation should be tailored to a subject familiar to management: business. Talk in terms of what the business objectives are and what the payoff to the business is in developing a specific application. Do not put in technology for technology's sake.

You need to develop an appreciation of what really does exist out in industry and what your organization's strategic needs are. This broadening of activity will enhance the overall success ratio.

To develop a plan that pulls the various needs together, you must:

- Define the application
- Determine the communications technologies involved
- Investigate all available products to meet the technological needs
- Present a business plan outlining the costs and benefits of the application to management
- Develop a pilot program to test application
- Test and evaluate the application
- If needed, modify and enhance the application
- Implement the technology on a production scale

The planning and communication result for the organization becomes rooted and acts as a vehicle for developing the plan to its greatest potential.

In addition to tactical issues, it is crucial to appreciate the strategic view. It is important to communicate effectively. Management personnel should feel they are not constrained by a lack of technological knowledge.

Planners will not be constrained if they know all the requirements. The department needs to come up with applications and do the necessary research, but it also should try to do more than just implement the technology.

A pilot program can be used once an application is approved and a vendor selected. A pilot can demonstrate that the vendor and the user are able to work together, not only to define the requirements, but to solve problems that come up during installation.

A pilot program allows the working group to verify that something works before the company attempts a full-blown implementation. Keeping management informed throughout the program's implementation phases makes them more aware of how communications technology applies to the company's plans.

Management recognizes that communications is a key to growth. Considerations such as how your particular planning applies, or if you are implementing a project soon enough to tie in with other projects, need to become a shared responsibility, with everyone involved.

One example of a team effort in creating strategic applications is First Wisconsin National Bank.* Without disrupting service, the bank replaced its entire data communications system in eighteen months and installed a private T-1 backbone network that serves more than 300 banking sites in four states.

After management gives approval to an initial zero-phase proposal, it is time to expand the proposal to include a complete assessment of the costs and benefits that will result from continuing the project.

One of the most important parts of an internal business proposal is a description of the proposed new system. It is here that the system's function, its scope, and its limitations are outlined.

It is wise to consider several important rules: First, you must thoroughly understand what type of information is needed. Next, you need to ask yourself the following questions:

- Should user knowledge of phone features or their satisfaction with them be tested?
- Should user preferences for future services or their perceptions of future needs be gauged?

These types of issues form the basic structure of the proposal.

The system's architecture, and the way that architecture relates to the current system architecture and interfaces, is discussed along with performance goals and ways of assessing performance. In addition to the more technical aspects of the new concept, any policy or organizational changes that might be necessary in implementing the system should be outlined.

The operational requirements of the new system should be laid out in the proposal. These include requirements such as power and the physical location of the new equipment. Include as much as possible about the system in a clear and understandable manner.

The detailed financial justification and cost evaluation for the project can be done with the help of the organization's accounting and finance department.

---

*Network World. December 22, 1983.

An 18-month planning process involved MIS marketing staff and customer service representatives. The result was a five-year plan for a T-1 and fiber network that matched a five-year plan for the bank, such as mainframe and disk storage capacity plans as well as plans to install workstations at bank locations.

The proposal should include a description of the personnel needed to implement and operate the system. It should include what user training will be offered and what plans will be implemented for converting to the new system. Security and disaster-recovery considerations must also be dealt with in the proposal.

Internal business proposals should include alternatives to the proposed plan and a realistic assessment of the technical and financial risk involved in implementing the plan. As it comes time to approach potential suppliers of the new system, an RFP must be prepared and submitted to vendors. Exactly when this part of the process occurs varies. A major capital expenditure might require an RFP first.

The acceptable tolerance for error on a major capital expenditure can be only 10 percent. Because you can only miss by 10 percent, you need to do an RFP to get the most accurate numbers.

There are four basic sections to an RFP. The first details the rules for submission. It lays out all the requirements for the proper submission of a vendor proposal. This includes the final data and hour of vendor submissions, evidence of the vendor's financial stability, and other customer stipulations.

In the introduction, the RFP should describe the user organization and the problem the RFP is seeking to address. The next section of the RFP covers system requirements. In some cases, it will list the average number of long distance and local calls made by each station. This quantitative data will allow vendors to build a model.

Maintenance considerations should be a part of the system requirements. They should be put into this section of the RFP. Here the user asks the vendors to describe the availability of spare parts for equipment.

Vendor response time to system problems and breakdowns is an important consideration and should be defined by the vendor. The vendor's definition of a major and minor failure should also be included. Some vendors can have backup PBXs on trailers ready to be loaded onto a plane and shipped out to you quickly in case of a fire or other disaster.

Another section of the RFP should relate to the facility and support requirements, including all the requirements of preparing a telephone equipment room. Heating, ventilating, and air conditioning, as well as power considerations, should be addressed.

The weight of the equipment and floor-loading considerations can be an important part of facility and support requirements. The conditions for acceptance of equipment should also be covered in a written agreement between user and vendor.

NETWORK PROFILES

In order to aid in the definition of proposals, a total of six profiles can be used to define the local network design criteria:

- The user profile
- The usage profile
- The geographic profile
- The applications profile
- The hardware profile
- The special requirements profile

These profiles allow you to formulate a description of the present and potential local network users and their environments. Most profiles should be prepared for a number of time periods: present, near future, and far future. Some information will remain the same, while other factors can be expected to change.

The user profile should include a listing of all the departments that use computer resources. This listing will include the number of people within each department, their job title, job description, and related information for each user group.

Written surveys can be used to determine the major characteristics of the user group. These surveys can be combined with structured meetings with the department heads of the groups.

The survey questions must be written clearly and edited carefully to weed out any biased wording that might invalidate answers. Should the survey be used to measure response intensity? In that case, closed-ended questions should have response scales such as strongly disagree, slightly disagree, neutral, slightly agree, and strongly agree.

A crucial part of survey-taking is distribution. The task is to define the target. When a total sample is impossible or impractical, the survey should employ statistical sampling techniques in order to draw valid conclusions about the entire population. The usage profile will outline how the current computer resources are used. This should include access to outside resources and peak periods.

The profile should include the listing of applications, programs, and terminal emulations in use. The peak periods for activities such as communications, printing, database inquiry, and file transfers should be listed. This can be done in terms of the number of transactions per day and per the peak hours, the hours of operation, file server input/output transactions, print transactions per type of printer, the number and size of records stored, and retention and access guidelines.

Another consideration is the system sensitivity to downtime (SSDT). This is a rating of the effect of downtime. A high SSDT indicates that downtime has a severe effect upon operations. The maximum acceptable response time (MART) of the network is also important in most applications.

The geographic profile is concerned with the physical distribution of local network users. This can be done for the present, the near future, the distant future, and the far distant future.

A code number may be used to describe specific departments and buildings in an installation. Each department or building can be subdivided further using codes.

The applications profile will outline the use of information-processing resources in completing the tasks of each department. Applications such as order fulfillment and payroll will have a real-time priority over applications such as human resources or fixed-assets management. The applications profile will also be concerned with how network downtime affects operations.

The hardware profile lists each user group and its computer hardware configurations. The lists will include the manufacturer, model, amount of main memory, amount of disk storage, operating system, tape backup facilities and software packages. Peripherals such as printers, plotters, digitizers, and high-resolution monitors will also be listed.

Disk drives should be categorized into family types. This should also be done for tape backup units.

Modems should be listed in terms of speed, Bell compatibility, and software compatibility. Other information that can be listed includes communications protocols and network applications to which access is needed. In addition to the hardware features, the network software environment must be considered. Such items as the support for file and record locking, remote access, print spooling, and the functions important for electronic mail might be listed.

The pace of technological changes is rapid in microcomputer packaging and performance. If we spend our time attempting to assimilate the effects of these developments, there will be little opportunity to focus on the particular application needs where small computers are needed.

You might ask how rational decisions can be made in the use of these products if the technology is moving too swiftly for measured analysis. One technique is to employ a freeze-frame approach as utilized in other specialties. Consider the communications capability of one system, viewing it from a relatively practical level.

NEEDS ASSESSMENT

The needs assessment is an important step in the process of selecting a network that is right for the needs of the user organization. The initial data collection is a part of the needs-assessment process. The data should be analyzed for patterns and trends.

The criteria to be used during the needs-assessment process depend on the requirements of the user community and the priorities of the needs that must be serviced by the network. The heaviest users should be identified and their needs examined in relation to the needs of the complete facility.

The needs assessment should address the specific networking requirements as well as the overall requirements of the system. These requirements will tend to drive the selection process and eventually become a part of the requests for vendor proposals.

The needs assessment should always include an analysis of business and organizational factors. The organizational factors might include the willingness to use current state-of-the-art technology, while business factors include the company's present and expected position within its industry.

In order to guard against installing an obsolete network the sources of future growth must be recognized and quantified. This growth must be realistically considered over the expected lifetime of the network.

The initial information gathering should be done using a list* similar to that shown below:

NETWORKING REQUIREMENTS

Organization: Date:

Person(s) contacted:

Position(s):   Telephone(s):

A. INTERFACES
  1. Organizations involved with in order to function.
     Organization • Building • Function
  2. Volume and frequency of paper flow between organizations.
     Organization • Drawings • Reports • Documents • Other
  3. Volume and frequency of data flow between organizations in units per time interval.
     Organization • Files/Rec. • Tapes • Disks • Other

B. COMPUTING EQUIPMENT CAPABILITIES/CAPACITIES
  1. Current network capability
  2. Maintenance
  3. Types of computers accessed
  4. Single terminal access
  5. Access through other computers
  6. Organizational sharing
  7. Sharing controls
  8. Size of organization
  9. Percent using terminals
  10. Types of user
  11. Types of terminal devices
      Device • Location • Quality • Baud Rate • Data Type
      • Dial/Dedicated

---

*Guidelines for the Selection of Local-Area Networks, by G. B. Ennis, L. J. Gardner, and M. K. Graham. (Washington, D.C.: National Bureau of Standards, July 1981). Pub. No. ICST/LANP-81-5

12. Types of dedicated or dial-up lines existing between computers
    Source • Destination • Baud rate • Dial/Dedicated
13. Utilization levels of each terminal/device
14. Database management system in use
15. Query language

C. SYSTEM OPERATIONS AND UTILIZATION
1. Operational sequence flow diagrams
2. Applications in use
3. Response-time requirements
4. Support-time requirements
5. Security requirements
6. Environment of physical equipment

D. EXPANSION AND GROWTH PROJECTIONS
1. History of past studies or reports on productivity improvements and system upgrades
2. Potential for automation
3. Automation requirements
4. Terminal/device requirements, present, and future
5. Types of devices needed/desired
6. Growth factors
7. Desired/expected productivity improvements
8. Future use/status of present equipment
9. Past growth patterns
10. Current problems
11. Desired improvements

E. UNIQUE CHARACTERISTICS OF THE ORGANIZATION

This information gathering will eventually be used as the basis for a more formal requirements document which will be used to identify the alternative solutions. The requirements document can be expanded or contracted as required. A typical outline for such a document is shown below:

REQUIREMENTS ASSESSMENT DOCUMENT OUTLINE

A. DESCRIPTION OF CURRENT METHOD OF OPERATION
B. BASIC REQUIREMENTS
1. Overview of New Operating Methods
2. Processing Logic
   a) General description
   b) Data flow
   c) Usage frequency and timing
C. DATA REQUIREMENTS
1. Data Storage Structure Layout

2. Data Dependency and Data Currency Requirements
3. Accessibility Requirements
4. Data Elements
   a) Identity
   b) Relationships
   c) Size
   d) Frequency of access
5. Peak Access Periods and Volumes
   a) Daily
   b) Weekly
   c) Monthly
   d) Yearly

D. OPERATOR INTERACTION REQUIREMENTS
   1. Remote access or control
   2. Geographic distribution
   3. Other interaction requirements

E. EXTERNAL SYSTEMS INTERFACES
   1. Data Volumes
      a) Average
      b) Peak
   2. Flow times
   3. Frequency
   4. Medium

F. PERFORMANCE REQUIREMENTS
   1. System availability
   2. Response time
   3. Data accessibility
   4. Frequency of operation
   5. Record retention duration
   6. Backup frequency
   7. Audit
   8. Security

G. SPECIAL OPERATIONAL REQUIREMENTS

H. GROWTH AND EXPANSION REQUIREMENTS
   1. Data storage
   2. External interfaces
   3. Number of terminals to be supported
   4. CPU throughput
   5. Geographic dispersion
   6. Estimated percentages
      a) 3 years
      b) 5 years

I. IMPLEMENTATION REQUIREMENTS
   1. Schedule

2. Equipment and human resources
3. Existing systems
4. Languages and operating systems
5. Software portability
6. Assumptions
7. Responsibilities
8. Conversion
9. Testing
10. Acceptance criteria

Another task is preparing an alternatives analysis report. This involves constructing a decision matrix and listing the advantages and disadvantages for each alternative solution. The requirements assessment outline can be used as a guideline in developing this report. The structure of a typical report is shown below:

ALTERNATIVES ANALYSIS REPORT OUTLINE

A. INTRODUCTION AND BACKGROUND
B. SUMMARY OF RECOMMENDATIONS
C. ALTERNATIVES CONSIDERED
    1. Assumptions
    2. Alternative descriptions and qualifications
D. TRADE-OFF STUDY RESULTS
    1. Requirements
    2. Weights
    3. Scoring alternatives
E. EXPECTED SYSTEM COSTS
F. RECOMMENDED SOLUTION
    1. Selection criteria/justification
    2. Preliminary system
        a) Definition
        b) Specifications
    3. Functional flow
    4. Expected expansion
G. PRELIMINARY SCHEDULE AND PLAN
    1. Phase I - system selection and preliminary design
    2. Phase II - system installation and testing
    3. Phase III - system operation and support

NETWORK TECHNOLOGY EVALUATION

The needs assessment should quantify the needs and requirements of the network for the present as well as during the next several years. The needs-assessment analysis should provide two important results. One is a clarification of the support requirements for the users, and the other is a prioritization of the individual needs that must be serviced by the network.

The next logical step is to evaluate which technology to implement. This requires an understanding of network technology. The viable communications solutions for most companies include using facilities provided by the telephone company and installing a new area network. The three basic types of area network technologies are:

- The private-branch exchange (PBX) for data-only or integrated voice/data
- Baseband signaling using twisted-pair wiring or coaxial cable
- Broadband signaling using coax or fiberoptic cable

The important technical considerations include:

- Configuration alternatives
- Types of computers supported
- Typical device speeds and maximum network speed
- Signaling medium and capacity per cable
- Maximum network distance
- Support for gateway and bridge processors
- Support for video and voice as well as data communications
- Installation

Table 2-2. International Standards Organization's Open-Systems Interconnect Model.

Layer 1	Application layer—user application process and management functions
Layer 2	Presentation layer—data interpretation, format, and code transformation
Layer 3	Session layer—administration and control of sessions between entities
Layer 4	Transport layer—transparent data transfer, end-to-end control
Layer 5	Network layer—routing, switching, segmenting, blocking, error recovery, and flow control (CCITT X.25)
Layer 6	Link layer—establish, maintain and release data links, error and flow control (bisync, HDLC)
Layer 7	Physical layer—electrical, mechanical, and functional control of data circuits (RS-232C, RS-449, CCITT X.21)

PBX TECHNOLOGY

PBX-based networks are typically star or tree configurations using in-place wiring as the signaling medium. These networks are often implemented in installations that have been renting telephone equipment. It is possible to provide support for integrated voice and data workstations. A variety of devices are available for purchase to support facility-owned PBX facilities.

The PBX acts like a large, automated switching matrix. It is similar in function to a telephone switchboard, except that the PBX does not switch the physical wire connections between users. Time-division sampling techniques are used to sample each conversation. The sample is then converted into a digital representation. By synchronizing the paths of the digital signals from the various extensions, the PBX establishes a virtual circuit connection.

A PBX network can be used with intelligent or standard terminals. Microcomputers can be connected to central databases using terminal-emulation software.

These microcomputers can also be used as integrated voice/data workstations. They can support voice, data and facsimile communications but generally not video. Each network usually cable supports one voice channel with control signals and one data channel.

A comprehensive cabling plan is important to allow expansion for any new installation. The added cost to the initial installation is small compared to the cost of pulling cabling each time a new device is added.

The PBX type of central point configuration is a source of total system failure, so many centralized products use a backup processor as well as electrical backup provisions. Others use a distributed architecture or redundant processors for each node. Voice/data network products are made by American Telecom, ATTIS, Ericsson, GTE, Harris, Intercom, Mitel, Northern Telecom, Nippon Electric, Rolm, and Siemens. Data-only network products are available from Develcon, Gandalf, and Micom.

The maximum network distance for microcomputers is 20 meters without modems or repeaters. A limited-distance dataset is normally used to connect the microcomputer to the PBX switch.

The cost per port for a typical voice/data PBX can range from several hundred to well over a thousand dollars each for voice and data transmission. The cost per port for a typical data-only PBX is about half as much.

The first PBXs basically mechanized the functions of human switchboard operators. Being mechanical in substance, they occupied a large amount of space. They also logged no usage statistic and handled only voice (analog transmission). They later used electronic switching and accommodated digital communications.

Many PBXs allow interterminal communications, electronic mail, peripheral device control, storage and forwarding of voice messages, and calls routing for cost-effectiveness. Others switch terminals between computers using the same techniques for call transfers between phone extensions.

AT&T, Northern Telecom, Inc., Mitel Corp., and Fujitsu Business Communications have introduced low-end PBXs designed to extend digital switching to users with under 100 lines. AT&T has introduced the System 75 XE, which is capable of handling 40 to 600 lines. Northern Telecom SL-1 ST is designed for users with as few as 32 or as many as 400 lines.

Mitel SX-50 is designed for the under 100-line user, and the Fujitsu Focus 196 addresses the needs of users with 50 to 150 lines.

Fujitsu offers a digital replacement of its analog Focus 100 for the 40- to 60-line PBX application. Most PBX vendors offer feature-rich, low-end PBXs. Fujitsu's Focus 196 PBX, for example, comes standard with an automatic-call-distribution system and station-message-detail recording.

Northern Telecom, Mitel, and Fujitsu offer custom features or different versions of their small switches.

Least-cost routing via private branch exchanges remains an important way for communications managers to save on telephone costs. In the late 1970s and early 1980s, complicated WATS banding schemes brought about the widespread use of PBX software programs that enabled users to access the most economical carrier for each call. Now, the interexchange carriers offer graduated calling options such as AT&T's Pro America II and III, megacom, and MCIs Prism.

Many of these new carrier options do not require the same kind of detailed analysis as the earlier WATS systems, which accounted for such items as geographic service bands and line packing.

Complicating the task of least-cost routing, at least 16 states allow long-distance carriers to compete with local telephone companies for toll traffic within local access and transport areas.

To effectively provide least-cost routing in today's environment, a PBX must be capable of using the area code of a call's intended destination, as well as the desired exchange. For intra-LATA long distance, the PBX must be able to use the first three digits of the area code, as well as the three-digit exchange.

Adding this six-digit screening is a function of both hardware and software because it requires both extra memory and processing power. Older switches tend to have limited memory. Most large PBXs built after 1975 provide the six-digit screening.

But even with newer and larger PBXs with adequate memory and the processing power capable of providing six-digit screening, there are additional tasks: obtaining the latest tariff information and updating the so-called lookup tables in the PBX least-cost routing software.

The long-distance carriers can supply information about tariff changes, but there is little incentive for them. In many cases, users must turn to consultants for both tariff information and help in actually updating the PBX tables. Intra-LATA long distance competition can produce marginal savings, and users in pursuit of such savings often need finely honed, least-cost routing capabilities.

The newer PBXs use many of the same microprocessors that today's microcomputers use. These microprocessors track which lines are overcrowded or underutilized and provide diagnostic test capabilities.

A twisted-pair cable is used for all devices, including telephones, data terminals, printers, microcomputers, and facsimile devices. Analog telephone sets are connected to a line card using one path for the analog voice signal and a separate path for the digital-signal control.

The microcomputers are usually attached to the time-division multiplexer (TDM) bus using an add-on data module. This TDM bus samples each line card or data module and aligns the time sample of two line cards or data modules to provide a communications path.

In its most common form, a PBX operates as a stand-alone private-branch exchange. It typically will support four types of trunk lines: a combination of incoming and outgoing trunks, direct inward dial trunks, wide area telephone service (WATS) trunks (and equivalents from common carriers other than AT&T), and tie-line trunks (usually two and four-wire leased lines supporting supervisory signaling). A fifth type of trunk interface for digital T1 (1.544 Mbps) channels is supported by some mid-range PBX switches. When functioning as a stand-alone PBX, most incoming calls are handled through an attendant console that greets the calling party and connects the call to the desired extension.

Multiple-attendant consoles are usually found in mid-range PBXs. The attendant console has the highest privilege level of any station attached to the PBX, including station and trunk status displays.

The attendant console can override a station as well as interrupt a connection already in progress. The attendant console usually is connected to the PBX using a 25-pair, twisted-pair cable.

Data traffic can be multiplexed in either the telephone set or at the line card in the PBX switch. When the telephone set has an internal digitizer, the voice is multiplexed with the data onto the telephone cable using time-division and two or three twisted-pair cables. The voice and data paths are separated at the switch.*

If the telephone station does not have a digitizer, the analog and digital signals travel through separate cable pairs from the telephone set to the switch. The analog voice signals are then digitized in the line card of the PBX switch.

Microcomputers connected in a PBX network can be switched to a local host computer, another local microcomputer acting as a file server, or an outward dial trunk on the public, switched-telephone network. A connection can also be made to an X.25 packet assembler/disassembler (PAD) gateway for connection to an X.25 public data network (PDN) such as Telenet or Tymnet.

---

*This technique is used in the Intercom IBX, the Northern Telecom SL-100, and the Rolm CBX-II.

The microcomputers all can share a leased line, with the X.25 PAD multiplexing several data streams. Remote microcomputers can be connected to a statistical multiplexer through a leased data line into a private branch exchange.

Low-end telephone equipment (fewer than 16 extensions with fewer than six trunk lines) is becoming a retail item. AT&T sells the product through Sears Roebuck & Co., which is competing with discount stores handling brands such as Panasonic. The cost per line of these low-end products is about $300, making it difficult for traditional suppliers to compete.

A mid-range PBX can begin with as few as 16 stations and extend to approximately 800 stations. A 16-station PBX is quite different from an 800-station PBX. The user must be able to extend the potential life of the product. The mid-range PBX can be linked to a network of PBXs and thus require the features of a larger switch.

Most PBX switches support asynchronous data traffic up to 19.2 kbps, which is a common requirement for microcomputers. Some support 56 kbps and NETBIOS communications of more than 1 Mbps.

Many PBX switches allow a direct connection to a T1 transmission line that operates at 1544 Mbps. The T1 link can be divided into voice-grade and data channels, which is very cost effective compared to the cost of individual voice-grade lines.

The AT&T System 85 uses T1 switching with pulse-code modulation (PCM) to create digitized voice data at 64 kbps. A maximum of 512 time slots is available for each processor, and a maximum of 16 processors can be supported. Separate voice and data signal paths are used for a per station cost of about $1000.

The Rolm CBX-II PBX uses a proprietary switching technique with a sampling rate of 12 kHz. The companding or compression technique is nonstandard, and proprietary telephone sets are needed in order to take advantage of the data-switching features. A data terminal interface (DTI) is also available as part of the telephone set. A proprietary submultiplexing scheme is also used in which several data streams are combined into a single voice path to reduce the requirements on the TDM bus. A protocol converter is available for use as an SNA gateway processor, and there is also an X.25 PAD gateway.

The Telecom SL-1 and the AT&T System 85 incorporate a port-doubling configuration in which voice and data traffic use separate cable pairs.

The Northern Telecom SL-1 uses a T1-compatible internal switching technique with 64 kilohertz sampling. Protocol converters are available for an X.25 PAD gateway and SNA gateway.

The Rolm and Northern Telecom PBX architectures are based upon hardware introduced in the late 1970s. Rolm and Intercom use true multiplexing techniques to combine voice and data traffic over the same two-cable pairs. The Intercom IBX allows direct digital connection to Bell's T1 transmission link and supports more than 4000 simultaneous conversations. Connection cost is approximately $1000 per connection when 25 percent of the voice stations also support data. The Rolm CBX-II and Northern Telecom SL-1 connection cost is about 10 percent for the same data mix. The AT&T and Intercom architectures are more recent and more expensive.

All manufacturers of mid-range PBX switches offer a proprietary line of digital telephone instruments with integrated data adapters. Data devices, including asynchronous ASCII terminals, are connected to the data port on the telephone instrument.

In-house equipment can be connected to PBX to allow local as well as remote users to connect to the facilities. IBM mainframes can be connected to the PBX by attaching data models to the front-end processor ports. The data modules typically are connected to ports on the station PBX cards. Minicomputer ports are connected in a similar way.

The largest selling non-Bell PBX is the Northern Telecom SL-1. Northern still markets this switch in configurations between 32 and 5000 stations. The switch architecture has remained basically the same except for the recent Meridian data-bus enhancement.

A private-branch exchange can also function as a stand-alone key telephone network. When functioning as a key telephone network, each individual telephone instrument must be able to access each trunk line used by the customer. Analog 2500-type telephone instruments are not supported because they do not have the capability to access multiple lines, and they cannot accept control signals used by the cards in the central PBX switch. A key telephone network does not have a console, because every telephone instrument has the same ability to access every trunk.

The PBX can be programmed so that only one station rings audibly while the remaining stations ring only visually; the audible ringing instrument then functions like a console. The designated attendant would answer incoming calls as they ring, put the calling party on hold, intercom the party being called, and then wait until someone picks up the held trunk as indicated by the console phone.

In the available AT&T configurations, small Merlin key service units support four trunks and 10 stations. The larger Merlin key service units support a range of configurations, from 10 trunks and 30 stations to 30 trunks and 70 stations. The digital Merlin equipment, called Merlin II, is available in configurations of between 8 and 72 stations. The basic station instruments include 5, 10, and 34 button models.

All stations use the same station card as well as the four-pair wiring. Each Merlin telephone includes a speaker but no microphone. The speaker can be used for voice announce and on-hook dialing.

The Merlin offers a high-quality voice, especially for conference calls. Every party is audible, even in six-way calls. Automatic route selection is available as well as call recording.

The digital Merlin II offers most of the functions available on the AT&T System 25 including asynchronous data switching up to 19.2 kbps. The Merlin product line includes a 24-trunk, 48-station key service unit.

The AT&T System 25 covers 24 to 200 stations with a maximum of 240 ports. The 25 employs a universal bus, universal card slots, and a universal addressing scheme. The control cards include a central-processing unit, memory, and service, which can be placed in any of three slots. The remaining slots are used for trunk and station cards.

The 25 uses erasable/programmable read-only memory (EPROM) chips on the control cards for programming. When the PBX is initially powered, it goes to a default configuration. Program upgrades are accomplished by unplugging memory chip packs and replacing them.

The 25 uses many of the same stations as the Merlin equipment. A user can employ the older Horizon 10-button MET instruments as well as analog 2500-type single-line instruments.

There are two types of station features: fixed and assignable. The fixed station features include transfer, conferencing, and hold. These features are available on the multiline instruments. Station features are programmed by the administrator and include speed dialing and call coverage. Station instruments include:

- A 34-button instrument with 32 programmable keys and a message-waiting indicator and speaker
- A 10-button instrument with 8 programmable keys
- A 5-button with 3 programmable keys
- A hands-free answer set with 10 buttons and a speakerphone

The 25 uses a 34-button instrument as the console. This station can handle from 20 to 200 stations with software features including return busy calls, return no-answer call, message-waiting indication, and call release.

The 25 supports only asynchronous data communications at speeds up to 19.2 kbps. The switch uses a chassis occupying about six cubic feet. The installed per station cost of a system is from $650 to $1000.

The advantages of PBX networks include proven technology and resource availability. They are the least expensive option for users with an existing telephone switch.

The repackaged AT&T System 75 XE is an expandable edition of its midrange PBX product line. It is intended for small and medium-sized businesses. The 75 XE is available in configurations of from 40 to 600 stations.

It can be configured for automatic call distribution. The 75 XE can be interfaced with AT&T's AUDIX voice-mail equipment. It supports direct-inward-dial trunks using trunk-interface cards. The 75 XE supports asynchronous data up to 19.2 kbps as well as synchronous data up to 56 kbps.

Data, control information, and digitized voice are combined in the digital telephone instrument. The traffic is controlled using AT&T's proprietary protocol, DCP (Distributed Communications Protocol), between the instrument and the 75 XE.

A 75 XE can operate as a node on an Information Systems Network (ISN) and as an end-point node on an Electronic Tandem Network (ETN). The 75 XE supports a DS-1 frame T1 interface and the digital multiplexed interface (DMI) for direct attachment to a minicomputer. Advantages of the 75 XE compared to predecessor models include lower startup costs, reduced pricing, and improved modularity.

PBX networks support data rates of from 300 bits per second (bps) to 9600 bps, with 56 kbps possible but rarely utilized. The disadvantages include limited device capacity, transfer rates, and bandwidth, which makes them undesirable for heavy volume applications. PBX applications with low speed and low-volume traffic include communicating word processors and copiers, microcomputers, interactive minicomputers, terminals, and printers.

PBX-based networks are adequate for moderate numbers of communicating microcomputers and asynchronous terminals. They are not as useful for data entry or synchronous data devices communicating in the 9.6 or 56 kbit/s range.

PBX networks can support message store-and-forward applications, protocol conversion, and connections to value-added X.25 packet data networks such as Telenet and Tymnet.

BASEBAND NETWORK TECHNOLOGY

Baseband networks employ signaling techniques in which each station sends digital signals over a communications path that is shared by all of the network stations. Each station will listen for messages and responds to those messages that are addressed to it. Baseband networks can use the star, ring or bus topologies.

These connections can use twisted-pair telephone cable, 50-ohm coaxial cable, or EtherNet coaxial cable. Twisted-pair telephone cable that is already in place will normally be implemented in a star configuration.

Baseband networks are often used for networking a small group of microcomputer workstations. The microcomputers are provided with baseband transceivers that plug into the expansion bus. A microprocessor on the board manages the network protocol.

These systems typically employ rates of from 2.4 to 56 kbps with a maximum network speed of 10 Mbps. They support one data channel with 10 to 100 devices sharing the channel by circulating the data, which can be in the form of packets.

The total transmission distance is typically less than 20 kilometers, with each individual segment separated by less than 1500 meters. The microcomputers are usually much closer to avoid the high cost of baseband repeaters.

Some baseband networks support digitized voice as well as data transmission. They do not often support analog facsimile or video transmission, but some support digital facsimile.

Among the first baseband networks was Datapoint's Attached Resource Computer (ARC), which has been in use since about 1975. Initially, the most common application was the connection of intelligent workstations.

These networks allowed the sharing of information among users that needed to be interactive, such as found in engineering design groups.

EtherNet cable-compatible components have been available from many sources including 3Com, Digital Equipment, Ungermann-Bass, and Xerox. Baseband local networks based upon twisted-pair trunk cables are also available from Corvus and others. The cost per port ranges from $500 to $1000 each. Baseband local networks allow a mix of different types of traffic and applications than is possible with PBX networks. A typical application is a cluster of microcomputers sharing a database and providing electronic-mail services.

Baseband local networks use time-division techniques to channel a single stream of high-speed digital bits among the users. A ring topology is often *used* with speeds ranging from 19.2 kbps to 10 Mbps.

The cost of baseband technology is similar to that of broadband, depending upon the data rates and station configuration. Baseband networks usually do not support digitized voice. They are limited in distance and have less growth capacity than broadband networks.

EtherNet trunk cables normally do not exceed 1500 meters without the use of a local bridge. Some baseband systems use twisted-pair wiring in combination with RG-62, RG-59, RG-11, or EtherNet coaxial cable.

Existing telephone wiring is normally not suitable because most baseband products require a ring configuration. Telephone wiring supports a star configuration which is rarely used for baseband. Baseband cabling costs are similar to those of broadband when the number of nodes exceeds 15.

Besides the network medium, the strengths and weaknesses of the different access methods must be considered. The basic types include token passing and the different versions of carrier-sense multiple access (CSMA), collision detection, and collision avoidance (CSMA/CD and CSMA/CA).

Carrier-sense multiple-access methods are best suited for networks with an aggregate network utilization of less than 30 percent and a terminal device utilization of 10 to 40 percent. Carrier-sense networks are suited for asynchronous terminals. They should not be used for synchronous terminals with use rates higher than 30 percent or for digitized voice applications when these are mixed with data traffic.

Token-passing techniques should be used for terminals with use rates of greater than 30 percent. Token passing is most useful when devices access at a regular rate. These applications include scientific/engineering workstations, real-time process control, factory automation, and some office automation applications. Token passing becomes saturated when a small group of devices needs to exchange data at high utilization rates. Brief discussions of some typical baseband network products follow.

Datapoint's ARCnet was designed for connecting microprocessor-based workstations. It operates at 2.5 megabits per second and can support up to 255 devices. Both bus and star topologies are supported. Token passing is used with an average price per user device of about $600.

The 3Com EtherSeries uses a bus topology and operates at 10 megabits per second with CSMA/CD access. It can support 1000 devices with an average price of about $600 per connection.

Digital Equipment's Decnet is an EtherNet baseband network that uses a tree topology. It supports up to 1024 user devices with a trunk channel speed of 10 megabits per second. CSMA/CD access method is used with protocol converters for X.25 and SNA/SDLC devices. The average connection cost is about $700.

The Corvus Systems Omninet network is a baseband network that uses twisted-pair cabling. It uses a bus topology and operates at 1 megabit per second using carrier-sense multiple access with collision avoidance. The network can support up to 63 users with an average connection of $800.

The IBM PC Cluster is a coaxial baseband network that uses CSMA/CA access. PC Cluster operates at a slow 375 kbps and supports up to 64 users. Station costs run about $900.

The Nestar Plan network uses RG-62 coax cabling connected in a tree topology. It operates at 2.5 megabits per second with token passing. Protocol converters for bisynchronous 3270 and SNA/SDLC are available. The average station price is about $600.

The Ungermann-Bass NET-ONE is a baseband system for IBM and compatibles. The interface is provided in the form of an expansion board. EtherNet cable or RG-59 cable is used in a linear bus topology. The network operates at 1.3 megabits per second and uses CSMA/CD access. The average price per user connection is $600.

The Orchid Technology PC net is also a baseband network for IBM and IBM-compatible microcomputers. The network uses RG-59 and RG-11 coax in a linear bus topology. CSMA/CD access is used with transmission at 1 Mbps. The average connection cost per user is $700.

Baseband implementation costs are often reasonable for moderate to small numbers of devices. Baseband networks support high data-transfer rates and are suitable for applications such as intelligent workstations that must share database information.

Supporting voice transmission can be difficult and is often costly compared to PBX. Intelligent terminal devices are required that constantly monitor the network for messages. Compared to broadband, baseband networks typically support lower-speed devices, have lower growth potential, and lower distance capabilities.

NOVELL NETWARE

The Novell NetWare/s-Net is a server that allocates disk sectors based upon microcomputer requests. This is done with a special version of the MS-DOS operating system. The network software tracks the opening of files and can restrict a second user if the file is already in use. Some file servers depend upon the application program as well as network hardware to prevent two users from simultaneously updating the same data.

The Novell NetWare includes a proprietary network as well as a software operating environment that is used on top of other vendors' local networks. Local-area network architectures from IBM, AT&T, and Apple are supported. The NetWare/S-Net is a star-based baseband network for connecting up to 24 microcomputers using dual twisted pair cable as shown in FIG. 2-1. The microcomputers are connected to a file server that employs the Motorola MC68000 microprocessor to control workstation access to the disk subsystem as well as printers. IBM PC, XT, and AT computers are supported as well as IBM compatibles and the Texas Instruments Professional. Each file server in the network can support up to 500/MB of storage and five printers.

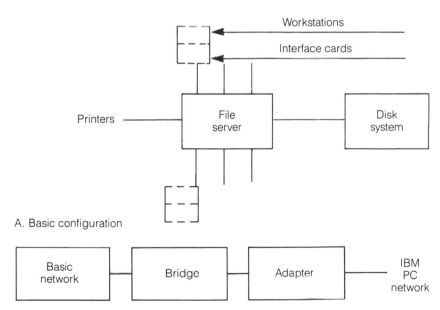

Fig. 2-1. Novell Netware/S-Net.

Multiple file servers can be connected to expand the network's size and geographic coverage. Each connected microcomputer requires an interface card. A true file-server configuration is used in which the microcomputers intercept an application program's service request to the server's hard disk as the workstations open, read, and close files. Other networks can interrupt only when they need to read and write to the server's hard disk.

The file server's hard disk is divided into directories which appear as individual hard disk drives to the user. A system manager assigns passwords to allow read/write or read-only access to the directories.

With the convergence of data processing, communications, and office automation, all vendors are striving to provide a more integrated solution. A network represents a commitment to a facilities architecture and a business strategy.

Part of the solution might be available only from different vendors because of the number of pieces that are interconnected. Difficulties can arise when competing technologies are nurtured in different parts of the same organization. The coexistence and connection of competing architectures is usually a trying experience.

Even as the communications industry continues with deregulation, many users opt to stay with AT&T products. AT&T has many products in place that they will promote as part of an integrated communications architecture that uses existing telephone wiring.

The AppleTalk network is a baseband network that uses a bus topology. It may be used for connections up to 1000 feet with twisted-pair wiring connecting up to 32 devices. The network operates at 230 kilobits per second using carrier-sense multiple access with collision avoidance (CSMA/CA).

The frame format is based on synchronous data-link control (SDLC). Connection cost is low, about 20 percent of the cost of an IBM PC-Net connection. This is because the interface hardware is a part of the communications built into each Macintosh, which includes a Zilog 8530 communications microprocessor with interface driver and receiver chips.

The AppleTalk protocol architecture has the following relations to the seven-layer open systems architecture of the International Standards Organization (see TABLE 2-3).

The physical layer uses shielded twisted-pair wiring with baseband signaling. The data-link layer uses the AppleTalk Link Access Protocol (ATLAP). Each node on the network has ATLAP facilities for packet assembly and disassembly.

Microcomputers on the network compete for access to the network bus through the link layer. Contention is resolved using CSMA/CA in which each device with data to send listens to the network. When a device is sending, the other units must wait until the channel is free.

In order to avoid data collisions, the following collision avoidance techniques are used. When a free channel is detected, each device must wait 400 microsec-

Table 2-3. AppleTalk Protocol Architecture.

ISO-OSI architecture	Apple Talk protocols	
Applications layer		
Presentation layer	Printer access protocol	Filing protocol
Session layer	Name binding protocol	
Transport layer	Routing table protocol	Transaction protocol
	(Socket-to-socket delivery, Internet)	
Network layer	Datagram delivery protocol	
	(Node-to-node delivery)	
Link layer	Link access protocol	
Physical layer	Media	

onds in addition to a random interval until a line reservation signal is sent. Each device is held, within these limits, until they are allowed to begin the collision-avoidance reservation cycle.

The CSMA/CA technique may become overloaded rather easily compared to other methods. It is the same access method used by IBM in its first local network, the PC cluster.

At the network layer a high-level protocol called Datagram Delivery Protocol (DDP) is used. It manages the transmission of packets among units called *sockets*. These are logical network units that link the local Macintosh applications to the networking functions. DDP manages the node-to-node message delivery on a signal network and the socket-to-socket delivery for the Internet connections.

The transport layer uses a routing maintenance protocol (RTMP) for locating the desired paths between nodes in the network. This protocol is integrated with a zone information protocol (ZIP) that is used to identify the location of nodes in the network.

The AppleTalk transport protocol (ATP) manages the delivery of packets between sockets, controls the packet sequence, and prevents packet loss. ATP employs a transmission identification frame for each packet. A data stream protocol is used to manage the full-duplex data transmission between sockets.

The fifth or session layer uses a Name Binding Protocol to control the translation between the object-oriented names of the network nodes and the numeric addresses used by the network.

At the presentation level, a printer-access protocol controls print-server functions, and a filing protocol controls file-server functions.

Automatic network addressing is used, so there is no need for a dedicated network control station. Each node has an eight-bit network address.

As a device is connected to the network, it transmits a network address bid. If there are no answers to this transmission, the originating device assumes the address. If another device answers the network address bid, a new address is calculated and transmitted.

AppleTalk is an inexpensive network for Macintosh microcomputers. Typical applications are those that mix high-resolution text with graphics using laser printers. Gateways are available for connection to other local network architectures such as the PC Network and EtherNet.

BROADBAND NETWORK TECHNOLOGY

Broadband networks support tree and bus configurations and use single or twin 75-ohm coaxial cable. They operate using device speeds of 2.4 kbps to 3.0 Mbps. The channel speed is typically 5 to 6 Mbps.

Broadband network components are based on community antenna television (CATV) technology that is relatively mature and widespread. The maximum geographic distance is 50 kilometers. Frequency-division techniques are used to allow the simultaneous transmission of many conversations on a single physical channel. The different device or user groups connected to the network are assigned to frequency ranges called *bands*. The transmission is analog, and digital signals require conversion using modems. The high bandwidth can support hundreds of voice, video, and data channels in the network.

High-volume and large-bandwidth applications are most suitable for broadband technology. Integrated voice and data communications, as well as video requirements, can be supported. The broadband networks with their large bandwidths and high transfer rates allow multiple data channels for data and video. Disadvantages include adding stations in dense areas, the analog signaling requirement, and the initial expense for network installation. Broadband is especially suitable to facilities with a number of buildings in a campus-like environment. The cost per port is approximately $700 to $1000 each. Components are available from a number of vendors, including 3M/Interactive Systems, Applitek, Codex, Concorde Data Systems, TRV, Sytek, Ungermann-Bass, and Wang.

Broadband local networks can be implemented with single and dual cables. Some network products, such as Ungermann-Bass's Net/One and Interactive Systems' LAN/1, are available in either a single or dual-cable configuration. Many are available only as either a single-cable implementation or as a dual-cable implementation. The dual-cable broadband products such as Wang's and Ungermann-Bass's usually are more expensive than single-cable broadband products such as Sytek's Localnet 20.* Dual-cable implementations, require two coaxial cables with two connections and two interface circuits per user. This increases the cost along with the additional space required for the cable and mounting hardware. The signals must be isolated in the terminal equipment, and additional time is required for cable installation.

---

*The IBM broadband network for microcomputers, PC-Net, is based upon the Sytek network technology.

A building that is wired with coaxial cable for the maximum capacity offers considerable savings compared to installing coaxial and twisted-pair cable on an as-needed basis. Broadband local networks can support most types of communications.

Broadband local networks offer the highest bandwidth capacity. They are also the most costly for an initial implementation. Over the longer term, broadband networks can be cost effective for applications with a high volume of voice, video, and data communications. They are also the only technology that can be used for full-motion, full-color, video transmission.

All broadband local networks require separate transmit and receive paths for bidirectional operation. The radio-frequency signals are sent through the trunk cable to the headed unit, which controls the signal distribution throughout the network. The *headend hardware* provides the required modulation, demodulation, addition, and frequency translation functions to the network.

Broadband networks support the T-1 data rates, which are generally used between Telco central offices. The data rates range from 9.6 kilobits per second to 20 megabits per second. The channels are normally divided into bands of 6 MHz. These in turn are often subdivided using channel-access schemes such as carrier-sense multiple access with collision detection (CSMA/CD) or carrier-sense multiple access with collision avoidance (CSMS/CA). Token passing is also used. This allows broadband local networks to support EtherNnet and other baseband configurations within a frequency band of the broadband configuration. The maximum throughput on the main data channel can range from 128 kilobits per second to 20 megabits per second with a maximum user device transmission rate of 19.2 kilobits per second to 2.5 megabits per second. A limit of 10,000, stations is possible with a maximum geographic distance of ten miles. Protocol conversion with a gateway processor is possible for conversion from asynchronous ASCII to either IBM 3270 bisynchronous protocol or X.25 bit synchronous high-level data-link control (HDLC) protocol.

Notable differences exist between broadband network products in their support for coax-attached terminals. Some only support IBM 3270-type CRTs, while others such as Wang support only Wang workstations.

They also are suited to those applications that require large bandwidth and high-volume communications such as the following:

- Integrated voice and data workstations.
- Real-time-critical response-interactive applications.
- Central-processor-to-central-processor file transfers.
- High-speed data workstations such as those needed for computer-aided design and computer-aided manufacturing (CAD/CAM) applications.
- High-volume, high-speed print stations including laser printers.
- Video applications.

The major disadvantages of broadband networks are the difficulty and cost of design, implementation, and expansion, especially in densely populated areas.

Broadband networks provide the bandwidth required to support total voice, video, data, and facsimile transmission. The cable and cable interface requirements are an expensive undertaking.

In environments with large volumes of data and some video requirements, such as universites and hospitals, they can be cost effective.

IBM's first local network was the PC Cluster. The PC Cluster was viewed as IBM's response to pressure for a local network product. This product was received as technically inadequate for most needs.

IBM's PC Network followed the poor response to the PC Cluster. IBM's PC Network is a broadband network that uses one frequency channel for the transmit signals and a second frequency channel for the receive signals. Each channel operates at 2 megabits per second. The network uses the carrier-sense multiple access with collision detection (CSMA/CD) technique, which is employed at the transport layer to control station access to the network.

IBM's most recent local network is the 4 Mbps Token Ring, which can operate on shielded, twisted-pair cable.

The first five layers of the seven-layer open systems interconnection (OSI) model starting from the physical layer through the session layer are implemented in the network adapter hardware. The PC adapter card is used, which contains a microprocessor, two communications processors, and 16 kB of random-access memory. The sixth and seventh layers of presentation and application are supported with the PC Network software.

The IBM PC Network allows IBM microcomputers to be networked together, with an AT acting as a shared file server as well as a stand-alone device. The hardware configuration is shown in FIG. 2-2. The 2-Mbps PC Network could use gateways between the PC Network and the Rolm CBX series. The PBX capabilities of the IBM-Rolm CBX products would provide a net-

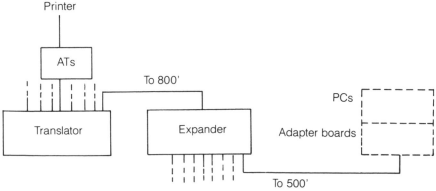

Fig. 2-2. IBM PC-Network.

worked PBX environment using the internal T1 capabilities of the CBX-II along with submultiplexing for data.

The PC Network hardware consists of the network adapter card for each microcomputer connected to the network, cabling and connectors, including base expanders, and a central translator unit that is required for each eight computers connected to the network. Eight microcomputers can be attached to this base unit, and up to 72 microcomputers can be connected using a maximum of eight base expander modules. Drop cables can be used in lengths of up to 200 feet for connecting the microcomputer to the expander module.

The total configuration can be extended to support up to 255 PCs. These can be spread over a radius of several miles in a custom-engineered implementation.

Devices that act as file servers or have shared printers attached to them must have a hard disk. Attached microcomputers use the network as a virtual hard disk. File sharing allows the PCs to segment and share the hard disk as well as a directory or file within a segment of the hard disk. The software required is MS-DOS 3.1 and the PC Network Control Program.

Network software includes password protection for shared files with read only and read/write access. The software will also control printers for which a user is authorized.

Most of the PCs connected to the network will no longer require printers or printer interface boards. Each PC has access to three printers without even one serial or parallel printer interface board.

Messaging programs are menu-driven and allow the user to display, file, or print messages. Message distribution lists can be stored, and messages can be forwarded to another user. The cost of the network interface hardware, including the broadband wiring, is easily justified in view of the potential savings. The main menu appears when the user logs on and indicates if there are any messages.

When a message is sent to a user, a status line at the bottom of the screen flashes. The user can temporarily suspend the application, handle the message, and then return to complete the suspended application.

Translator costs are approximately $600, and the network adapter cards cost about $700 each. A mix of 20 microcomputers can be networked at a cost of approximately $20,000. Based on a three-year life without depreciation, this equates to about $1.25 per user per working day. Factoring in the benefits of accelerated depreciation, it drops this to about 60 to 70 cents per day. This does not include maintenance, administration, and support of the network. Typically, network product life cycles can be expected to run from three to five years.

There are other benefits that result from the productivity gains to be had from information sharing and the new applications that are possible in the network environment.

The PC Network was developed as a low-end LAN for users that need to connect a limited number of PCs and communicate at speeds of no more than 2

Mbps. Token-ring technology offers more capability and fits well with SNA. The token-ring network will run on the IBM cabling system. The initial version was available at 4 Mbps using shielded twisted-pair wiring, and future versions of up to 16 Mbps are possible using metallic cable. Fiberoptic versions could be capable of achieving speeds of up to 100 Mbps. This could be used for transmitting voice via SNA protocols.

The IBM Token-Ring Network is a token-passing broadband network that includes a 3274 cluster controller to support peer-to-peer communications and plug-in communications cards for the PC. Both the controller and the card use a 5-chip token ring parts set that Texas Instruments developed for IBM.

Other vendors can incorporate these chips in compatible products. Proteon, for example, offers its 10-Mbps ProNet LAN, which is compatible with the token-ring.

IBM's token-ring development work started in the early '70s at its research labs in Zurich. A token-ring network was demonstrated in Geneva in 1983. The prototype network interface units were originally as large as a PC.

When an XT is used as the server, the number of nodes is limited to about ten. An AT can support 16 to 20 nodes. An APPC/PC applications interface is available that implements the LU 6.2 subset protocols for program-to-program communications. This interface allows users to develop programs for peer-to-peer communications.

An asynchronous communications server program is available for low-speed communications over switched lines. The program runs on a PC, XT, and AT in a nondedicated mode and supports two dial lines per server. Multiple asynchronous servers are supported on the Token-Ring.

Communications out of the ring may be through the Rolm CBX II, a PBX, or public switched network using modems. The program manages outbound calls and accepts incoming calls for asynchronous communications programs running on the PCs. It requires exclusive use of the asynchronous communications adapters and can limit the performance of other programs running on the same PC.

A Token-Ring-to-PC Network program is available that uses a dedicated AT or XT for running the interconnect program. Network applications communicating across the link use NETBIOS. The program provides limited access to nodes between the two networks.

A PC Network SNA 3270 Emulation Program can also be used to provide access to host applications for 3278 Model 2 devices. The PC running this program must be connected to protocol converters such as the 3708, 3710, or 7171, or to the Rolm CBX II using a Rolm DataCom module for the interface. The CBX II link supports data rates up to 9.6 kbps, while the protocol converters can operate at rates up to 19.2 kbps.

Access to 3270 devices is available in several different configurations. A stand-alone configuration is offered in which a PC is attached with an SNA/

SDLC line to a host. This configuration emulates the 3274 controller functions with a 3278 or 3279 terminal and 3287 printer.

A gateway can be used as a communication server for up to 32 PCs on the Token-Ring or PC Network using an SNA/SDLC line to the host. In a network station configuration, each station acts as a communications node on either network by emulating a 3278 or 3279 terminal and a 3287 printer. A gateway is used to access the host. A combined gateway and network station configuration can be used where the functions of the 3274 controller, 3278 and 3279 terminals and the 3287 printer are emulated.

NETBIOS is also available for a growing number of applications. The addressing scheme is simple to use, but because it was not developed for internetworking, the number of nodes that can communicate across a gateway is limited. It also does not communicate directly with mainframes.

APPC/PC tends to be less efficient as an applications interface, but it allows peer-to-peer communications between applications running on S/370 (with CICS), S/36, S/38, Series/1, and PCs. It can operate interactively or in a background mode, and it is supported on both the IBM Token-Ring and SDLC adapters.

In building an IBM Token-Ring Network, the low-cost, short-term solution is to use type 3 wiring because the twisted-pair telephone wire might already exist in the building. Two pairs are required for data, one pair for voice and one pair for 3270 traffic. Type 3 cabling limits the number of nodes on the network to 72. The distance between wire closets and workstations is limited to 330 feet, or 150 feet if the two wire closets are on the ring. The distance limits are based on the signal quality falling off as the distance increases while the noise level remains the same.

A media filter must be used to attach a PC with the IBM Token-Ring adapter to the type-3 media. This filter reduces the amount of signal radiation. It allows the IBM Token-Ring to support 4 Mbps data rates by eliminating frequencies that could radiate between pairs. A jumper cable is used to connect the Multistation Access Unit (MAU) to the type-3 wire.

The MAUs can be used to link up to eight nodes on the network, and MAUs can be connected together to expand the ring. MAUs can be installed in a 19 inch rack in a wiring closet.

Type-3 cabling is limited to supporting 4 Mbps. Types 1 and 2 are designed for higher speed data transfers. An 8-foot attachment cable connects the PC to type 1 or 2 cabling. For connecting two wire closets on the same token ring, type 1 coax must be used. Type 1 and type-2 (twinax) cable cannot be mixed.

Another adapter is available for 3270 traffic on type-3 cabling. The 3270 Coax-to-Twisted-Pair Adapter (CTPA) is a combination of balun and media filter.

The applications interfaces perform the functions of layers three through five in the OSI model. They are implemented in software rather than in ROM on the adapter card, as in the case of the PC Network. Implementing the applications protocols in software provides more power and flexibility for incorporating protocols such as SNA.

Two low-level and two high-level programming interfaces are available. The lower-level interfaces are for IEEE 802.2 data-link control and IEEE 802.5 direct physical control. The higher-level interfaces are NETBIOS and APPC/PC, which can be run concurrently on the IBM Token-Ring without an adapter card for each node.

PACKET TECHNOLOGY

In a relatively short time, packet switching has become an attractive alternative to leased telephone and data links. A leased line is dedicated to a single user, but packet switching allows a number of data users to share a high-speed link. The concept allows fast, error-free, reliable, and low-cost communications.

The operation of a packet network is similar in some ways to a railroad system. Each group of information is placed in a container (the railroad car) and carried to the next station. The information is then taken out of the first container and placed into another that will take it to the next station or node. The process continues until the information arrives at the final destination.

An actual packet-switching network transmits information in small groups. This is usually less than 128 characters to minimize the end-to-end delay. Many small-volume users can simultaneously share the facility, which results in a lower cost per bit of information transmitted. The user does not need to be concerned with the task of network operation or the cost of support. Message switching is not usually performed in real time because this would require large storage facilities at each node. A packet assembler/disassembler (PAD) is needed because before the blocks of characters are sent, the control and destination information must be added and the packet assembled. At the destination this information is removed as the packet is disassembled. The control and destination bits are then discarded, and the data is sent to the receiving terminal or other device. The PAD can be owned by either the user or the network. The PAD is similar in some ways to a statistical multiplexer because it allows terminals and devices to share one leased phone line for network access.

In a packet network, the user pays for the phone line from the device to the nearest node. A fee is assessed for each packet sent and the connection time.

The connection to a packet network requires the user to connect to the nearest node in the network using a dedicated or leased phone line. Then the originator sends a call request packet to the call port. This call request will force the network to assign a channel number that identifies the data and the address to which the data is routed. The connection is automatic and transparent to the user terminals.

The packet network is not sensitive to the transmission error problems that occur in dial-up networks. Error checking is done at each node to assure that the message is received correctly.

The packet network is less vulnerable to line failure than a leased data link. If a link should fail, the packets can be rerouted on other links.

Packet switching is a flexible data-communications technique, and like other shared systems, it is not necessary for the user to design, implement, and support the network linking geographically separate locations.

The AT&T ISN is an integrated, packet-controlled concept directed at users that are likely to have the necessary twisted-pair wiring already in place. FIGURE 2-3 illustrates the possible connections in a typical ISN (Information Systems Network) configuration.

The network uses twisted-pair cable to connect terminals, microcomputers, host computer ports, print servers and file servers to a Packet Controller. The Packet Controller acts as a data packet switch. Processing up to 48,000 packets per second, it controls and manages the traffic flow in the network. Terminal devices can be connected to a *concentrator*, which multiplexes up to 40 devices into a 8.6 Mbps data stream. The concentrators are connected to the Packet Controllers, which can be separated by up to 1 kilometer using fiberoptic cable. The Packet Controller can be connected to an AT&T System 85 or System 75 PBX through the 8.6-Mbps fiber data stream.

The network uses AT&T's Premises Distribution System (PDS) as the signaling medium. PDS can support voice, data, video, and facsimile. Fiberoptic cable can be used in addition to twisted pairs. PDS is adaptable to most PBX switch products. Most PBXs also will operate using the IBM Cabling System with type-2 cable. The cost per station is about $100 to $150 for twisted-pair cable and $150 to $300 for the combination of twisted pair and fiberoptic cables.

Asynchronous communications is supported with the 7404 Digital Voice Terminal, which allows simultaneous voice and data transmission through a digital port on the System 75/85 PBX.

Data modules are available that allow a user to switch IBM 3270 data messages from asynchronous terminals to IBM-compatible host computers using the System 75/85 PBX. These data modules range in price from $800 to $1600.

An adapter on the ISN packet controller provides support of IBM synchronous communications. 3270 BSC and 3270 SDLC cluster controllers can be used at speeds of 9600 bps and 19,200 bps. Another adapter provides for direct connection of 3270 terminals. Keyboard commands can be used to switch cluster controller ports between applications and host computers.

A balun interfaces the 3270 device's coaxial cable to a single twisted-pair copper cable using a 2.36-Mbps data rate. This permits movement without expensive cabling changes.

AT&T offers other local networking besides ISN with its 8 megabits per second of throughput in a star topology. STARLAN is a 1-megabit-per-second network based upon carrier-sense multiple access with collision detection

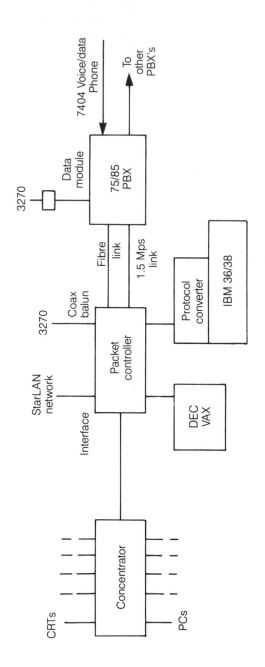

Fig. 2-3. AT&T ISD.

(CSMA/CD) access techniques. STARLAN is a local implementation that emulates Microsoft's MS-Net and the MS-DOS version 3.1 operating environment.

EtherNet, the industry standard baseband local network with a channel speed of 10 Mbps, is also offered along with STARLAN. Synchronous support for IBM's Systems Network Architecture and a high-speed transmission trunk for connecting ISN Packet Controllers as remote nodes is also available. Networks with more than 75 device prices have an average cost per connection of about $400 for asynchronous devices, $700 for STARLAN attached devices, and $1200 for IBM synchronous devices.

The data modules connect 3270 devices to a System 75/85 PBX by first converting the 2.36 Mbps data stream to 64 kbps. The 64 kbps data stream is switched through the PBX in a manner similar to other circuit-based services.

The protocol used between the data module and the PBX is the Digital Communications Protocol (DCP), a protocol that is proprietary to AT&T. PBX networking is possible through the Electronic Tandem Network (ETN) and Distributed Communications Systems (DCS) products. Gateways can be used to connect to a STARLAN or EtherNet local network.

An important feature is the ability to build networks by interconnecting an ISN with a System 75 or System 85 PBX as well as additional remote ISN environments. However, ISN should support T1 and Dataphone Digital Service (DDS) transmission channels for effective remote networking.

The use of the Digital Multiplexed Interface (DMI) will allow multiline host interfaces to be reduced to a single, high-speed, 1.544-Mbps data stream. Networking with the DMI interface and T1 transmission links can be an effective method for both local and remote requirements.

Several differences are apparent in the areas of premise wiring and office communications architecture between AT&T and IBM. Both offer twisted-pair wiring with a fiber option, but AT&T uses the traditional four-pair per station of non-shielded, 22 to 26 gauge wire, while IBM uses nonstandard shielded twisted-pair cabling for data and four pairs of conventional telephone wiring in their type-2 cabling.

3

Network Management

As network technology advances telecommunications, management has become more complicated. There are sophisticated tools to help, but there is much more information to deal with.

Decisions with far-reaching ramifications must be made in an environment that is changing faster than ever. The management of a voice network consists of much more than adding a line when users complain about a busy network. Integrating voice and data has become one of the major concerns.

Data networks have grown. Years ago there were only point-to-point circuits, including multidrop. Managing today's networks is difficult for traditional voice managers. A new way of thinking has to be applied.

As markets grow, they require more service. Thus, there is a need to consider the whole network, voice traffic as well as data traffic. All communications managers struggle to develop long-range strategic plans coping with the effects of rapid change.

Uncertainty about corporate strategy, the cost of telephone services, and future regulatory changes make planning difficult. For example, the Federal Communications Commission has proposed lifting rate-of-return legislation for AT&T and the regional Bell holding companies.

While it is unclear what the impact on service prices will be if rate-of-return regulation is eliminated, users must still make decisions today about the networks of tomorrow.

Other unknowns, such as the fate of the local economy and potential tax-law changes, further complicate the planning process. For example, a proposed tax on all interstate long-distance calls originating in Texas has caused some Texas firms to consider establishing a network switching facility in Oklahoma to avoid the tax.

While planning is more difficult in an uncertain environment, users know that it is also more important. Rather than trying to foresee every possible change, the best a communications manager can do is focus on meeting the company's needs. For example, for a company with very heavy data communications needs and heavy and equally private line voice needs, T-1 makes a lot of sense no matter what happens with rate-of-return regulation.

Voice and data networks are managed and measured differently. Statistics on response time, throughput, and error rate are required. Voice requires statistics on the traffic carried and the quality of transmission. Network modeling tools are available for this, and many of them are microcomputer based.

Only a few organizations have the sophisticated tools and techniques in place to give a true picture of a large voice network or a combined voice and data network. It can be difficult to justify the expense without first using the tool or demonstrating the technique. Gathering the data to run these tools can also be a major task and acts as a deterrent to getting the tools and techniques in place.

Poisson and Erlang modeling are often inadequate for newer networks in which queueing, alternate routing, and the extended holding times of data calls must be considered. Newer tools are required. Networks designed using Poisson analysis might work, but because they are over-engineered, the network has far more capacity than needed.

Many corporations are combining voice and data network management. The separation between the two becomes smaller as we move toward T1/DS1-based digital voice and data networks. Rather than a simplification of the network management tasks, this indicates, in most cases, the only really workable solution.

Today's network management usually includes Telex, TWX, facsimile, teleconferencing, and electronic mail. Data centers might use IBM hosts that have connections provided by the telephone companies for communications, but many companies are trying to get these mainframe communications onto other formats. The manufacturing organization might use Hewlett-Packard equipment. MS-DOS-based microcomputers and Apple's (used for graphics) also can be spread throughout the organization.

Outlying major data centers need network connections between them, this equipment should be integrated into the packet network.

The use of integrated networks, such as Integrated Services Digital Network (ISDN) also brings a new area of functions. These networks require a shift in thinking. Users will want end-to-end service, and it will not just be an applications or line problem. Problem situations must be handled from inception to completion without delay.

In an integrated network, voice and data networks share digital facilities, and combine voice and data management. Integrated also has other meanings. Integrated transmission facilities pipe analog voice and data and digital voice and data on the same routes. Digital voice, data, and image are multiplexed on the same trunks and are ultimately packeted at the switches. Integrated equipment includes digital switches and digital transmission facilities together with network management equipment and user interfaces all designed and implemented as a whole.

Integrated management structure means integrated tools and techniques and the required organization to go with them. Communications systems cannot grow or be improved without strong advocacy on the part of communications managers. In any of these processes, there must be a way to measure what is going on in any portion of the network at any given time. There also must be the means to process this information for modeling purposes, the means to predict the effects of changes, reconfigurations, and the ability to perform what-if analysis.

To the user, a terminal must appear directly connected to an application. The network must handle all the interim equipment, switching facilities, and connections without the user's notice.

Voice network management is illustrated in FIG. 3-1. This voice network management structure is typical for many organizations. Most of the work is done in the areas of implementation and operations together with the administration functions.

Network design must consist of more than an annual review of the telephone company study. This is often received months after the data is collected and often has questionable value. Other design functions include the addition or deletion of locations from the network and the addition of trunks to compensate for too many busy signals. Overcapacity can be identified and dealt with through the use of call detail recording (CDR).

ECONOMIC JUSTIFICATION

Telecommunications professionals must show sound economic reasons for making changes to the existing communications system. Major changes in communications systems generally fall into three catagories. The most common is system growth. Very often, organizational expansion requires more private-branch exchange stations, terminals, or other associated net hardware.

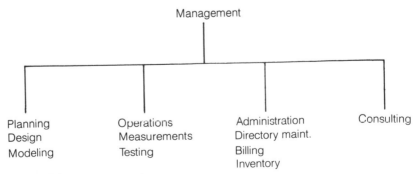

Fig. 3-1. Voice management.

System growth is most easily understood by upper management and likely to meet with quick approval when it can be directly attributed to personnel growth.

Problems are most often encountered when the need for a relatively small number of additional terminals or phones necessitates expansion of the entire system.

In the case of a PBX, for example, a relatively simple request for station equipment can mean that the number of PBX line cards and cabinets must be expanded. Suddenly it appears that the addition of ten phones will cost $50,000 because of the need for overall system expansion. This makes the addition harder to support. Managers have to illustrate the relationship between the additional company personnel and the need to expand the system.

The second major reason for a significant network change, a desire to improve performance, can also be tough to sell. The argument is generally based on system efficiency. For example, there might be a need to enhance network diagnostic software so that technicians can more quickly detect and resolve network problems.

The third and usually most challenging reason for making a major change in the communications system is the introduction of new technology designed to provide a strategic business advantage. This is usually based on the assumption that specific business problems can be solved or that productivity can be enhanced.

VOICE MANAGEMENT

FIGURE 3-2 shows the shells of a voice-management model. This view of voice management illustrates the links in the model and the forces that influence them.

In the center is the network management team. The first surrounding shell shows areas of direct responsibility. The next layer shows the main source of influence on the functional areas. The carriers are still the major force in this shell. Local telephone companies can limit operations and choices.

The final layer can be the most important. It shows the information sources that are used to generate plans and statistics to manage networks. The ability to get data is often limited, with most of it coming from call recording or the carriers.

Many organizations do not use alarm and traffic data from PBXs. Most switches generate the data, but unless users draw the data from the switch, it can be lost. Some switches generate large amounts of data, but unless provisions are made to capture it, the data is dumped or often overwritten.

For small networks, CDR/SMDR (call detail recording/station message detail recording) can be processed in-house, perhaps on a microcomputer. For larger networks, more processing power is needed, and data collection becomes a major task. A service bureau can prove to be effective.

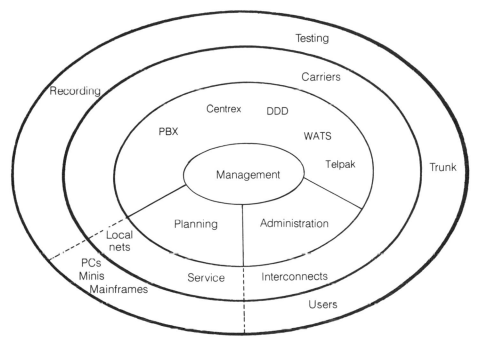

Fig. 3-2. Voice-management model.

More often it is necessary to analyze the CDR output for network-planning purposes. It can reveal the amount of long-distance traffic to a particular location.

Analysis of call records can show tie-trunk (network) usage. The problems involved with collecting data on voice management illustrate the problems with network management as a whole.

The response from upper management might be that we can get along without all the new bells and whistles. Comments such as these indicate that the cost of network downtime has not been well presented.

One can, for example, argue the case for voice mail, electronic mail, or a new wideband data application. The argument must present the advantages of the new technology in terms of additional cost and expected benefits. These cases require careful analysis and documentation, and such cases can be very rewarding for those who succeed in leading their organizations to use technology for strategic advantage. In instances where the expected benefits do not materialize, managers must carefully analyze why. Communications managers should capitalize on their victories and learn from their mistakes. Only then will upper management continue to approve major network changes.

The information is also collected in small bits. An automated method of gathering the data and generating reports is required.

The outer shell shows testing, which is limited in many cases. Without a good testing scheme, users become dependent on local telephone companies' routine maintenance. Users of large networks often experience having some portion of the network out of service, suggesting that maintenance is not rigorous.

Maintenance should be proactive, not reactive. The ability to quickly and economically test facilities on a regular basis results in greater reliability, fewer trouble reports, and happier users. Testing ability is often limited. Most PBXs can perform limited trunk testing and take faulty trunks out of service and print a report. This might be programmed to occur at specified times. Such tests are normally limited to the capabilities of the PBX.

More effective testing can be achieved using stand-alone, minicomputer-based devices. These can conduct tests on both digital and analog facilities as well as processing alarms. The drawbacks to minicomputer-based devices are their expense and their external relation to a switch. There is also the need for automatic trouble-ticket generation and management reports.

Even small networks can benefit from the automatic generation of electronic trouble tickets. These tickets contain some analysis, such as correlating alarms to possible trouble sources. These tickets require attention. They can be forwarded electronically to follow a contingency plan in case of problems. Upon final problem resolution, all parties are notified. Templates for routine operations might suffice, but the ability to create new ones must also be provided.

DATA MANAGEMENT

The data-management model (FIG. 3-3) should also take into account that many organizations are not set up with a centralized communications management department with a special data center and a separate data communications group. The basic features and functions depend on how an organization is structured, recognizing that there will have to be some consolidation of communications and computing. In the idealized model, the network management group sits in the center of the organizational structure, surrounded by users and the equipment for which they are responsible. The administration function is strong and among the forces of influence on the network, the carriers have a major role. The vendors also have an important role. Unlike voice, if there is a problem with some communications gear, the user's impulse is to call the vendor and not the carrier involved.

Users should formulate a plan based on a best-case view of the future telecommunications environment. They will be able to negotiate best with the carriers who seek to offer services designed with their interests in mind, not those of the users.

Much communications related software is available and most can be used to good effect, but it is often limited and can place further limitations on the net-

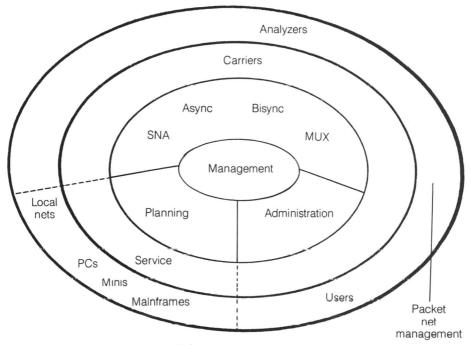

Fig. 3-3. Data-management model.

work. An intelligent modem, for example, is intrinsically an analog-to-digital converter and cannot be used on pure digital facilities. Much software is only effective for a given protocol. Packet-switched networks are an exception. They do not handle all protocols, but for those that do, management is excellent and the tools effective. Line loading, switch load, error statistics, alarms reporting, and alternate routing are available. Detailed operational and management reports are also available.

The essentials of managing both voice and data include knowing a number of factors relating to the health of the network in real time. These are:

- Alarm monitoring and reporting
- Line and trunk testing, both automatic and manual
- Throughput and response time statistics
- Line and switch loads
- Meaningful management reports and metrics
- Automatic trouble-ticket generation
- Electronic databases of inventory, operating in real time
- Active electronic directories
- Billing processes

To complete the process, the following requirements for a network management are necessary:

- Viewing the overall network in graphical form
- Zooming to a particular node or line component
- Separate voice from data at any node
- Viewing into all networks
- Analyzing CDR and other data to provide what-if analysis and other statistics
- Collection and processing data regardless of format

Without these considerations, network management is executed in an uncoordinated fashion. Voice networks are managed with a combination of carrier-provided studies, manipulations, and computer programs. Some vendors have stand-alone hardware/software devices that allow modeling, but do not permit integrated data and voice capability. Usually the voice is modeled and the data added later.

Preventive maintenance and testing should have established guidelines or standards. It should not be necessary for the user to indicate a problem exists, as is often the case with voice. This situation can be made better with voice management and measurement. The question of testing involves many aspects. Users must choose whether to buy additional equipment, use PBX facilities, or try some other method. Pure data networks can be better than voice, but there are still limitations. One is how to make the best use of expensive communications facilities. Existing tools for packet networks are an exception. Neither the carriers nor the vendors are much help. When response times start to rise, it is hard to determine if it is the application, the front-end processor, the line, the cluster controller, or a combination of these that is at fault.

Monitoring might require commissioning a study from a carrier or consultant, or purchasing line-monitoring equipment and developing the means to process the collected data.

Powerful and progressively more integrated tools are appearing. They are, however, not all things to all people.

A combination of available tools can be employed using a modular implementation scheme. This makes it possible to trade in portions as newer technology becomes available.

It is not only a question of changing the tools; it is also necessary to examine management philosophy. The distinction between voice and data is blurring. As the voice and digital worlds merge, the differences between the computing group and the communications group become less and less apparent.

Integration will take place on a much broader scale than it has in the past. In the future, dynamic networks should offer bandwidth on demand. These networks, based on T1 carriers, will use 56-kbps streams and share arrangements between voice and data. Multiplexers will be heavily used. Both 56-kbps lines and T1 cross-connects are in use by the carriers. They will also be used in many private networks.

Networks will evolve to combinations of T1 links between major locations, smaller groups of digital 56 kbps circuits for smaller locations, and single often multiplexed digital circuits, or analog circuits for light traffic.

The new management organization can be visualized as being totally responsible for all transmission facilities. This might not always be practical, but the overall function is needed. If a separate group cannot be justified, it might be possible to create an initial group using staff from the other major groups such as voice, data communications, and MIS (management information systems). They could be maintained within their original structures.

Transmission facilities management then appears to act as a common carrier or retailer of bandwidth. If the voice needs are such that two extra lines are required between locations, an order is placed with the new management group. It establishes lead times based on its relationship with the carriers, and provides service at predetermined and published rates within the company. This is shown in TABLE 3-1 as the relational links between the in-house common carrier, transmission facilities management, and the rest of the functional model.

This new group in effect owns the digital facilities, the digital cross-connects, and the multiplexers. Costs for these items are built into the rates to users. A certain amount of overhead is needed in terms of spare capacity.

With this system in place, the processing, for a new user might take the following form:

1. User requires access to an application.
2. User contacts application services at data center.
3. Application services verifies validity of request.
4. Application services contacts transmission facilities and requests connection.

Table 3-1. Functional Model of TFM.

User Services **Problem Management**	
Data networks	Transmission
Voice networks	facilities
Packet-switched networks	management
Local networks	
Video and teleconferencing services	
Metrics and modeling	
Other	

5. Transmission inputs data into network model and assigns bandwidth, updates inventory, revises spare capacity, advises on cost, and adjust data-center billing.
6. Application services issues user I.D., logon, and password and advises of user start date.

Problem management for a voice network problem could be handled as shown below. Assume that the problem was not found with centralized diagnostics.

1. A user identifies there is a problem.
2. A user calls the local contact such as an internal operator.
3. The local contact does a first review to determine if the problem is just a misdial or a true problem and generates a trouble ticket.
4. If the problem seems to be with the local switch, the local contact calls the local telephone company and copies the trouble ticket to voice services for their information.
5. If the problem appears to be network-related, the trouble ticket is transferred to voice services.
6. Voice services uses the available tools to determine if the problem is with PBX hardware or software. It deals with the problem, or if it is network-related, it transfers the trouble ticket to the transmission-facilities group.

With this system there is operational independence and responsibility in the individual groups. The voice group, for example, handles PBX software and hardware, table updating, and billing, and might have some responsibility for network design in terms of queueing time, alternate services such as WATS, and grade of service (the number of busy signals in a given hour). These would be handled as orders to transmission facilities that will be responsible for the overall performance and provisioning of the network. Sophisticated design tools might be needed to design the network based on voice parameters with a data overlay.

Integrated voice and data design requires many what-if scenarios to establish the viabilities of alternate routing, variations in queueing times, and the effects of long data calls.

Consolidation into digital networks will result in a single interface point for network testing. The type of device to be used for testing might be the existing network management equipment common to data communications. The electronic matrix switch has its abilities to monitor, test, and control the attached circuits. Other information is available from a PBX. This type of traffic data and network performance statistics can be sufficient to form the basic structure of the network tools. Automated trouble ticketing would also be a part of this.

A newer level of design programs are data network modelers, which are appearing along with integrated voice and data modelers. Rather than buy the entire package, it can be more advantageous to purchase packages of limited functions and contract for a study annually to calibrate a less sophisticated tool. This method has the advantage of gaining up-to-date measurements at least once a year.

The relevance of up-to-date information to model networks and their worth in giving meaningful data in terms of management reports and statistics is high. Many telecommunications groups are designing voice networks based on manipulations of call-recording data. Some microcomputer programs are even available using Erland B or similar formulas, but these cannot effectively model most mixed networks. The problem can be one of calibration. Existing formulas give a reasonable prediction at a given level of performance. If it can be established that the grade of service is equivalent to two busy signals out of every 100 calls, then the program might be able to give a reasonable indication of what effect on the line it will have.

In other cases, the program might indicate base-line performance level is in error. The program might not have the sophistication to deal with certain realities. These include, for example, tandem networks, where, when reaching a busy signal, a number of users hang up and don't try again, a number do try again, some number might be queued, and some might be switched to alternative services.

TWO-MOMENT MODELS

Recent developments in network modeling include two-moment models. In the past, most programs have been single-moment models, which means that they use one algorithm to model all conditions. They can use approximations for conditions subject to overflow or queueing. The two-moment model has more than one algorithm and is able to change its algorithm when conditions change. These programs are more able to model accurately a network for various levels of queueing, overflow, and alternate routing. This can be done iteratively.

Savings of up to 30 percent on network costs are possible with the more accurate network models. Also available are single-node optimizer programs that can save expenses on individual PBXs.

The cost of such services is not small, but it can usually be justified annually. It has the advantage of keeping networks up-to-date, maintaining the highest possible service levels, and minimizing service costs. The output of the total network study can be used to calibrate a micro or minicomputer-based tool on a link-by-link basis. The calibrated tool can be used to adjust the network with a high degree of confidence, perhaps monthly.

All the desired network-management functions should be available at a single location, not necessarily a single physical terminal, but on a single network—perhaps a central minicomputer base with remote collection and with computing devices at each network node.

NETWORK-MANAGEMENT CONSOLES

Network management can require a number of consoles because in a large network multiple problems often exist simultaneously. Each console would have access to all functions available to network management. A member of the network staff should be able to sit at any terminal, switch from one communications mode such as voice or data, to another, and window between applications.

Operators should be able to view the composite network (FIG. 3-4) and window to trouble-ticket information selectively, zero in on a node, and then separate out the voice or data connections.

The network management architecture (FIG. 3-4) will tie together all the fragments. Because of its modularity, it should accommodate planned growth.

The components are computer-based gear with hardware and software that will collect and manipulate data and monitor alarms for a small number of switches. Other versions will also be able to handle the traffic data generated by PBXs. Some devices will handle alarms and traffic reports for large voice networks and have a similar capability for data networks. These will be interconnectable and will provide the basis for full-scale network management. The end result is to provide integrated network management with some intelligent functions and the flexibility to adjust for changing conditions.

COMMUNICATIONS ARCHITECTURE

After identifying the organization's communications needs, the next step in planning is to develop a communications architecture. An architecture is a general scheme for meeting the organization's needs without always identifying particular hardware or service providers. It can allow vendors to be selected according to such criteria as price and demonstrated reliability.

Developing a flexible network architecture is central to communications planning. Because of rapid changes in technology and in corporate business needs, five years is generally the optimal life span for a major network.

OSI AND NETWORK MODELS

Information architectures are the key to more complete integration of current and future computing environments. The International Standards Organization's Open Systems Interconnection (ISO/OSI) reference model has done much to popularize the use of architecture in information networks.

An architecture can be defined as the set of rules necessary to build an object or to interface one object with another. An object can be many things. The ISO/OSI model uses a layered architecture to define a common array of communications protocols.

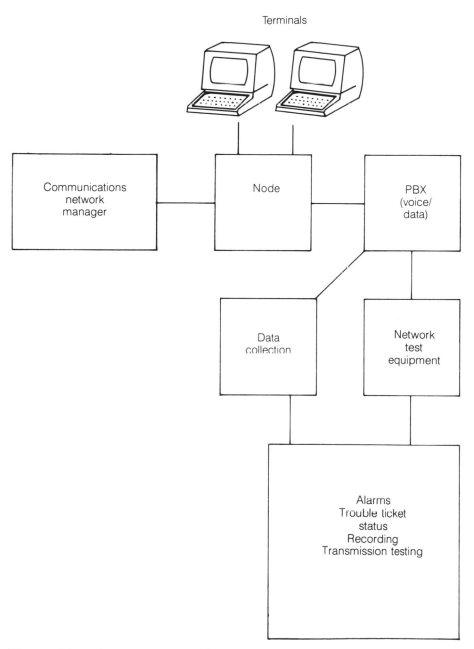

Fig. 3-4. Network-management architecture.

Confusion in the use of architecture results from architecture being used interchangeably with a standard. ISO has developed an architecture and a standard, and OSI is sometimes assumed to be both.

A standard is a convention established from a number of alternatives. Organizations adopt standards or conventions for information technologies to manage their spread, use, and integration.

These standards can already exist in an organization, or they can be adopted from de facto standards set by a majority of users or from formal standards set by standards committees. A standard can be established from a number of alternative architectures.

The ISO/OSI reference model and the protocols used to implement the model have been developed to address the need for information interchange between heterogeneous computing environments. The OSI reference model does not address the actual computing hardware or operations required to support information services, only the exchange of data between a network's two nodes. The ways that the data is accessed, obtained, and manipulated, and how the information is displayed, are outside of OSI.

There are other architectures that address the same requirements as the ISO/OSI model. The Transport Control Protocol/Internet Protocol (TCP/IP) and the Xerox Network System (XNS) are architectures that also treat the problems addressed by the ISO/OSI model.

Architecture management involves the evaluation of the architectures available in order to recommend certain ones as standards. The integration problem found in many computer networks stems from the hundreds of mainly proprietary architectures available, for example, IBM's System/370 and DEC's Decnet. The challenge of architecture management is to reduce the number of required architectures to a set of standards.

The process of determining which architectures to use as standards is similar to the process of selecting hardware or software products. Evaluation criteria can be developed based on requirements, goals, and objectives. Candidate architectures can then be evaluated against these criteria.

There might even be a make or buy decision involved, and if an existing architecture does not provide the capabilities needed, a new one might be developed. The MAP/TOP architecture developed by General Motors and Boeing is an example of such a development because existing architectures did not meet anticipated needs.

The development of a unified approach should involve communications managers, MIS department managers, and users in the adoption of a singular, organization-wide information strategy. Not only can such a plan reduce the organization's expenditures on information resources, but it enables the organization to make better strategic decisions about information handling.

Organizations need a common model that managers, analysts, and users can employ that will allow them to define network requirements with specificity and the compatibility to make integration easier. The terms model is used as a description or analogy to help visualize that which cannot be directly observed.

This model should be based on the concept of managing the processing and communications resources including the types of data, specific products, as well as the architectures.

THE SIM MODEL

While a number of different network-planning models (at different levels) have been proposed, the Strategic Information Management (SIM) model has been found to be useful for defining interactions and interrelationships.

To understand the SIM model, it is useful to examine the major differences and similarities between SIM and the International Organization for Standardization's Open Systems Interconnection (ISO/OSI) architecture and protocols. The SIM model addresses communications and other issues in a way that is similar to OSI.

The SIM model uses layers of different functions. These layers allow the establishment of protocol boundaries that define how information flows from one section to another. The Strategic Information Management model also uses modularity, so the replacement of a product in one layer has minimal or no impact on products in other layers.

In the SIM model, the layered concept is applied across all of the information components, not just the communications portion covered in OSI.

The SIM model consists of four layers: the user interface layer, the application layer, the data layer, and the processing layer. FIGURE 3-5 shows the ordering of the layers and the service categories found in each layer. Each layer can

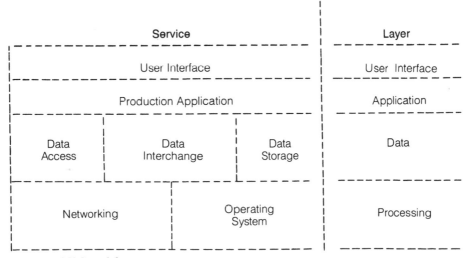

Fig. 3-5. SIM model.

use the services of any other layer at any time. The four layers comprise a complete representation of information resources.

The relationship between the SIM model layers can be illustrated by the metaphor of a vehicle. The various layers of the SIM model are analogous to the steering, fuel system, and engine that make a vehicle operational. The driver of the vehicle is not as concerned with the operation of its components as with reaching a destination. It is up to the automobile designers and mechanics to ensure that each component functions correctly and compatibly with the others necessary for the vehicle's operation.

The SIM user interface layer is concerned with presenting information to the user, providing the user with a view of the processes. The user interface layer contains the applications with the capability to handle the user's interactions with any information resource.

In the vehicle analogy, the user interface layer addresses the steering, the gas, and the brake pedals needed for operation. It also covers the information output, measurements such as the speed of the vehicle, and the amount of fuel available. The other layers of the SIM model can be treated in a similar way. Depending on the function of the production application, it might employ a user-interface package, a data-access package, and a communications package to perform a single task.

The data layer consists of support applications that manipulate data. The layer is divided into data-access services, data-interchange services, and data-storage services. The combination represents whatever is needed to manage data flow through the network.

This layer contains the fuel used by the vehicle in the analogy. The data layer holds the fuel lines for accessing the storage tank, the methods of storage, and the delivery system.

In the information model, the data layer includes the underlying database structure devoted to storage. It also includes the data models used to represent the data's type and location. The relationships between the data models and the rest of the model provide the link between the physical data and the applications.

The bottom layer is concerned with processing that is connected primarily by the interfaces between the rest of the information-processing applications and the hardware.

It contains operating systems and networking services, and it is the layer in this model that is concerned with the network's hardware and transmission equipment.

In the vehicle metaphor, the processing layer includes the engine and the interfaces between the engine and the rest of the vehicle. The engine provides the power to make the vehicle work. If the vehicle had multiple engines, the processing layer would need to contain facilities to coordinate each engine's operation.

The operating system services of the processing layer are the boundary services between the hardware and the rest of the resources. Operating system support software can address performance measurement, maintenance utilities, user libraries, and device drivers.

Networking services provide the communications between any two applications. This layer includes local-area networks, host-to-host networks, and network management.

Networking services and operating-system services are in the processing layer because they provide many of the same interface services. Each has functions concerned with the sending and receiving of data within one computer or between different machines.

Each SIM layer addresses a component required for successful operation. Each layer is important in the way that an automobile's engine is just as important as it wheels: removing either would make the vehicle useless. Layering allows the interfaces to be defined so that one engine can be replaced by another to improve performance.

The SIM model demonstrates how the components of each layer work with respect to each other. FIGURE 3-6 illustrates the relationships between the typical applications required to interconnect a workstation and a host computer. This model could be used to represent interactions between any two or more computers.

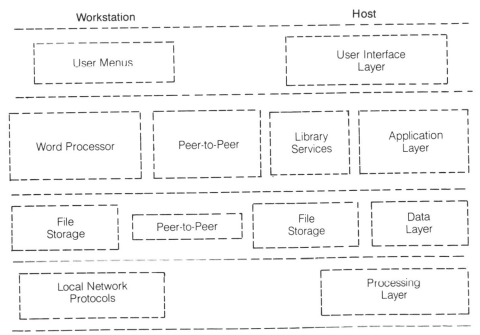

Fig. 3-6. SIM model for workstation and host applications.

The left and the right sections represent the workstation and the host. Each section has been divided into the four layers of a SIM mode. Developing the model for each computer will show the hardware connectivity requirements and the integration necessary for both the software and the data.

Employing the SIM model can be done in a number of steps. The steps might take place as follows:

1. Determine the level of detail. The content of the SIM model depends on the intended audience. A business-oriented view should minimize detail, while highlighting concerns of functionality, data integration, and ease of use. To maintain the same level of detail, ask if any object is a type or form of another object.
2. Formulate a view that addresses only the user interface and the application and data layers of the model. This step is not concerned with hardware or any of the elements of the processing layer.
3. Add the application functions. The detail functions in the application layer are derived from an analysis of the tools and capabilities required to perform each of the individual processes required. For example, these might include electronic mail and project planning. Other application functions can be related directly to one or more of the desired tasks.
4. Add the user interface. In most environments, the users desire a common user interface.
5. Add the data. The next step is to determine that data requirements for each of the functions in step 3. The need for data integration determines which storage units are grouped together.
6. Connect the components. The lines between components indicate the relationships between each of the components. Each line represents an information flow between separate functions. FIGURE 3-7 shows a typical corporate-wide application.

One of the strengths of the SIM model is the ability to determine if there are any problems in the design. Each function of the application layer should match a capability found in one of the layers of the conceptual view.

The specific flow of tasks can be evaluated for ease of use, functionality, and integration. SIM models promote consistency in the use of information technology. The resources required for a particular solution and their application can be displayed. Using a common model provides a link between processes and conceptual design. SIM provides a visibility for new concepts between different disciplines. The building-block approach allows each layer to be viewed individually with varying degrees of complexity. If the impact of the global information resource must be determined, the model is capable of an enterprise-wide analysis.

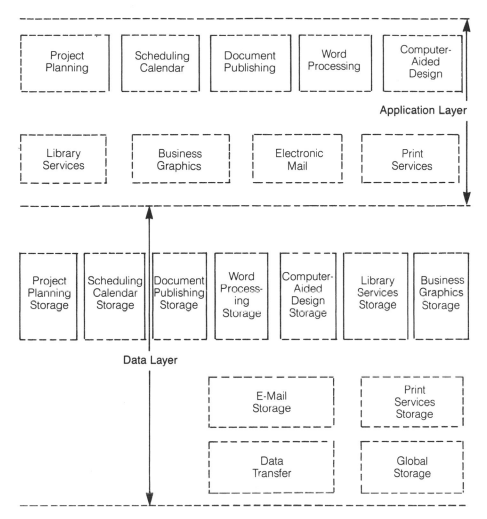

Fig. 3-7. Application and data layers for a typical corporate-wide SIM model.

Many forces can influence the effective design and delivery of information services. Resource modeling is the foundation SIM. The SIM model can be applied throughout the development life cycle of a network. The SIM model can be sketched easily or automated, and it helps the participants visualize how the effects of change in one phase of development might affect the project.

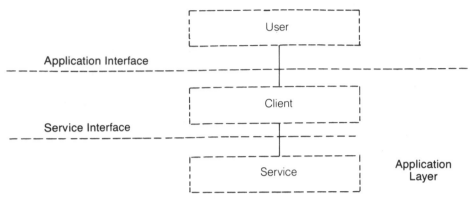

Fig. 3-8. DSF model nondistributed environment.

THE DSF MODEL

The ECMA* Distributed Services Framework (DSF) is a general model for distributed services. The combination of distributed services constitutes an integrated information service. Three layers: user, client, and service make up the distributed services framework. The model recognizes that each of the various distributed services communicate with each other as standards are developed.

FIGURE 3-8 show a nondistributed environment where the production and support applications reside on the same computational node.

If an application requires a distributed solution, a service interface that included communications protocols would represent the communications requirements to the remote machine.

The ECMA model requires that the distribution of functions be transparent to the user. This implies that the application interface cannot change. FIGURE 3-9 show how the client portion of the application remains with the user at the application interface.

The application interface has not changed. The service portion of the application has been distributed to another node, but the application interface remains the same. The service interface and the service-access protocols perform the separation. The client and the service are complementary and communicate through service-access protocols.

NETWORK-MANAGEMENT SOFTWARE

Truly integrated network control would enable users to manage a mix of vendors' data and voice-communications network equipment from a single point.

---

*European Computer Manufacturers Association

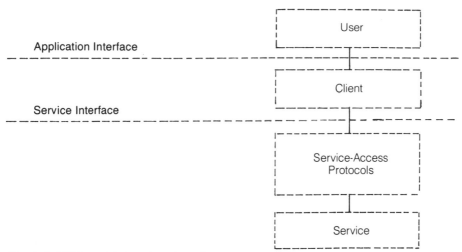

Fig. 3-9. DSF model distributed environment.

Many communications-equipment vendors with equipment ranging from multiplexers to matrix and packet-switching units have incorporated network-management schemes into their products. In some cases, this arrangement offers conflicting or incomplete diagnoses.

Several vendors produce packages that integrate the output of various network-management utilities, but there is little or no uniformity in the field. The International Organization for Standardization has yet to incorporate management as part of its Open Systems Interconnection reference model. In the absence of international standards, IBM might have set the de facto network-management guidelines. Both NetView and NetView/PC have received industry support from third-party vendors. This network-management software pair would not be the first IBM "strategic products" that failed. Some feel that the mainframe is the wrong locus for network management and control because of its limited view.

While NetView lays a groundwork for multivendor network management, the fact that it is a host-based software system might help its future as a comprehensive network-control facility. If NetView on the host is to control non-SNA equipment, it will require not only product-specific interfaces for NetView/PC, but an application program on the IBM host where NetView resides.

If vendors cannot supply that host-based software, NetView could become a one-way reporting system. It will collect information from non-SNA devices equipped with the proper interface, but will not be able to control those devices.

NETVIEW CHARACTERISTICS

IBM's NetView and NetView/PC provide a basis for network management. NetView/PC allows users to manage token-ring networks either locally, from a central NetView mainframe or from a remote console (FIG. 3-10).

The IBM Token-Ring Manager is a program that assists the user in problem determination and error recovery on the token-ring LAN. It can operate as a stand-alone application, or using such NetView/PC applications as alert-forwarding to the host and remote console support, can provide management for the entire network. Network errors are recorded in an event log. Other network events, such as stations joining and withdrawing from the network, can be recorded in the log to assist the user in problem determination.

Logged events can be displayed and printed for a specified station or period of time. Alert conditions are signaled to the operator using audible alarms and highlighted indicators on the display. At the same time, an alert message can be transmitted to NetView in the host.

NetView/PC's remote-console support allows an operator at a remote PC to use the local token-ring network manager. The remote unit communicates

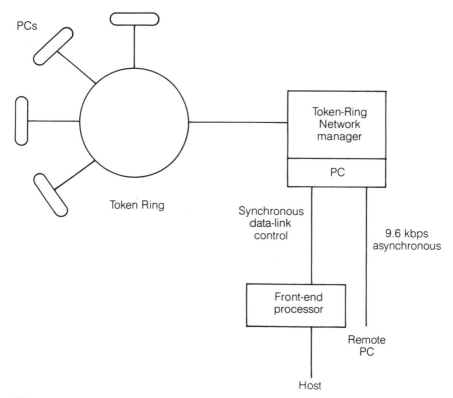

Fig. 3-10. Token-Ring management.

with the local PC over an asynchronous ASCII link. NetView extends Systems Network Architecture (SNA) management to information networks that include non-SNA, voice, and non-IBM products.

The IBM NetView Personal Computer (PC) program allows users to construct networks containing token-ring local-area networks (LANs), private branch exchanges (PBXs), or high-speed multiplexers. The PC version of NetView can be used for stand-alone network management with no host connection, for enhancing management based on NetView in the host, or both.

Multitasking support handles all services, including the disk-operating-system (DOS) partition, as separate tasks. Some tasks, such as external communications, require a large number of polling interrupts and processing cycles. These are offloaded to a coprocessor board, making more cycles and storage available for user applications running alongside NetView/PC.

Network error support is provided for three types of network error information: alerts, problem records, and service reminders. Alert-management software handles the alerts as they are passed to NetView from the applications. An alert manager timestamps alerts as they are received from applications and logs them to alert files. The software also maintains probable-cause panels to indicate the most likely cause of failure.

NetView uses a subset of SNA alert formats called Network Management Vector Transport formats. These alerts include such information as alert origin, severity, probable cause, actions recommended to the operator, and supporting data. Alerts are stored in a database.

The operator can define how each category of alert is to be processed, stored, and displayed. The problem records are created automatically when an alert is received or when the operator updates the problem log. Problem-management elements include: documentation of the recovery or bypass procedure used; assignments made to staff responsible for problem resolutions; tracking of status until the problem is solved; and sorting of problem records for statistical analysis purposes.

OPERATOR FACILITIES

The service-reminder functions provide additional backup for problem resolution. Either the operator or an application program can enter a service reminder with the time and data specified.

Operator services supported by NetView include control of the operator's interactions with the applications or with NetView, log ons at the remote console, and a help function. Other services are also available. An operator can interact with one application in the foreground while NetView continues to process requests for other applications when they are dispatched by the host or outside PBX.

A help facility gives operators at the PC interactive assistance and reference material. Help modules show how to use NetView, how to define alert-management criteria and how to manage files. Access-method services can format data from applications to a disk for input to PC-based report-generation programs.

The remote console permits an operator at one NetView console to assume control of another linked to it by an asynchronous communication line. To be effective, programs like NetView must support a broad range of communication products and protocols for connection to the host. The following communications methods are supported mainly by the coprocessor board:

- Half-duplex SNA synchronous data-link control (SDLC) is used to communicate with the host.
- Asynchronous ASCII devices, non-SNA equipment, PBXs, and remote-consoles. Options include off or even parity, one or two stop bits, half or full-duplex modes, echoplex, and data rates to 9.6 kbps.
- Switched-line support for auto-dial or auto-answer capabilities, with either synchronous or asynchronous transmission.

With its application-program interface, NetView/PC can offer the user a generalized implementation of the service-point concept. Services are provided to applications executing in a DOS partition managed by NetView/PC. Service requests use a call to one of the DOS assembler subroutines for alerts, operator communications, host data, and service-point commands.

NETVIEW AND SNA

In every copy of VTAM software running in a mainframe host computer, there resides a Systems Services Control Point (SSCP) function. (See FIG. 3-11) The SSCP acts as the network supervisor for the network components controlled by a particular host computer, referred to as that host's *domain*. In this way, the SSCP provides network-management functions for the terminals, controllers, front-end processors, and links in a domain, as well as for the host itself.

SNA networks with multiple host computers can contain multiple SSCPs, with each host controlling its own portion of the network or domain.

In order for the SSCP to manage network resources, the host must be able to communicate with all network nodes in its domain. The SSCP must be able to issue network management commands to other SNA network products and collect information about network status from remote nodes.

The SNA network products must be able to respond to network management commands issued by VTAM.* The function in each node that provides

---

*See *Systems Network Architecture Format and Protocol Reference Manual: Management Services* (IBM document No. SC30-3346) and *Systems Network Architecture Format and Protocol Reference Manual: Architectural Logic* (IBM document No. SC30-3112).

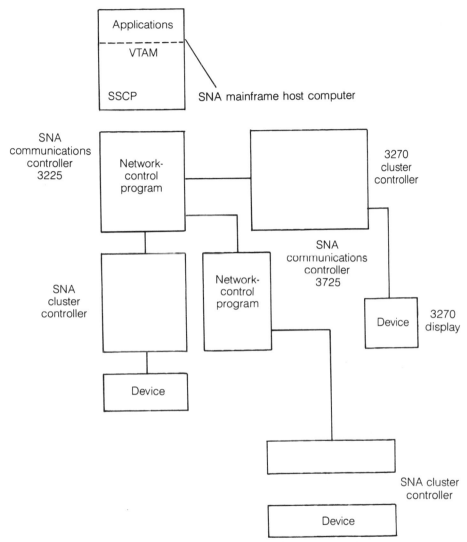

Fig. 3-11. Systems Network Architecture network management.

network-management capabilities is called a *physical unit* and is implemented in that node's software.

Each SNA node (host computer, communications controller or terminal cluster controller) has a built-in PU function. The SSCP communicates with PUs using a session (the logical connection) to the PU of each node in its domain.

The host VTAM-resident SSCP becomes the master for network-management functions, while the PUs in each node act as the slaves. Sessions between the SSCP and remote node PUs allow the centralized collection of network-management information in the host computer.

IBM also refers to VTAM SSCP hosts as *focal points* and other SNA node products containing PUs as *entry points* for network-management information.

NETWORK-MANAGEMENT COMMANDS

Network-management commands include a query command called *record-formatted maintenance statistics* (RECFMS). RECFMS can be sent between a remote terminal cluster controller (for example, a 3274) PU and VTAN to determine engineering change levels in the controller hardware, software or microcode, or to find the number of synchronous data-link control frames received in error.

IBM's generic network-management command or message structure can be used by the SSCP in VTAM to issue commands or to gather network status information. Network Management Vector Transport (NMVT) uses an extendable hierarchical encoding scheme made up of a major vector key, subvectors, and subfields.

The Response-Time Monitor lets VTAM collect response-time information from remote 3270 terminal cluster controllers. Different NMVT commands implement most of the common network-management functions.

NetView runs in an IBM host as a VTAM network-management application program. It provides a more cohesive set of SNA network-management tools in a single product (see FIG. 3-12).

NetView in the host works in conjunction with the SSCP function in VTAM to support network-management functions. NetView replaces and includes most of the following functions:

- Network Communications Control Facility
- Network Logical Data Manager (NLDM)
- Network Problem-Determination Application
- VTAM Node-Control Application
- Network Management Productivity Facility

NetView/PC runs on an IBM AT or XT and extends SNA centralized-network-management facilities to non-SNA environments. NetView/PC allows IBM or third-party application programs running in the Personal Computer to pass network management and control information, received from non-SNA products, to NetView running in an IBM mainframe.

NetView/PC can receive network-management messages from the Token-Ring Network Manager or from Rolm CBX Alert Monitor programs running in the Personal Computer, and it can pass the information to NetView in the host. NetView/PC uses an SSCP-to-PU session with VTAM to pass control information to NetView.

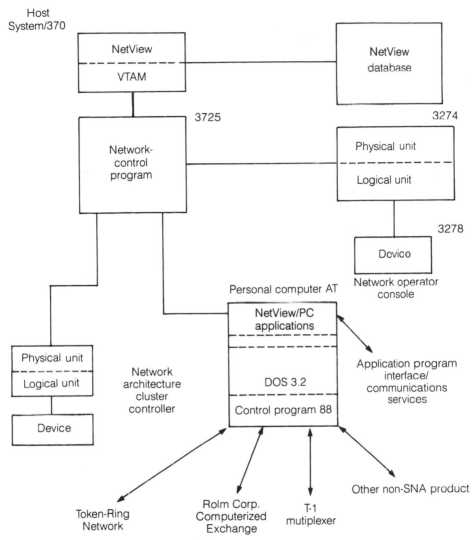

Fig. 3-12. NetView and NetView/PC.

Under IBM terminology, NetView/PC in support of non-SNA systems is a network-management service point passing network management information to the NetView focal point.

For large SNA networks, defining the network to VTAM (creating the VTAM lists) can be time-consuming and difficult.

For large SNA networks in which multiple 9370s are used as distributed processors connected to mainframes, NetView can provide remote-system console support. This permits 9370 operations, including network-management functions, to be performed by a single remote central site, minimizing the number of network-management workers required.

For smaller SNA networks supporting 9370 communications, NetView can be used with the NetView Network Definer. This permits operators of 9370s running VM/SP to define their network configurations to VTAM in an interactive fashion using menus and prompts.

COPROCESSING HARDWARE

The Real-Time Interface Coprocessor helps to keep the NetView/PC microcomputer from bogging down. It handles communications functions for a number of protocols, such as logical unit 6.2*, physical unit 2.0, and the synchronous data-link control. The board includes a microprocessor, memory, and software.

The NetView PC must provide its functions within many storage and performance constraints. Host communications require logical unit (LU) 6.2 sessions for data files and physical unit (PU) 2.0 sessions for alert processing. This requirement can be met with the Advanced Program-to-Program Communications for Personal Computer (APPC/PC) program. However, APPC cannot operate with a PC's limited memory together with enough code for NetView/PC functions. The solution was to offload all communication software, such as the LU 6.2, PU 2.0, synchronous data-link control (SDLC), and asynchronous protocols, onto a separate coprocessor. The Real-Time Interface Coprocessor (ARTIC) card, developed by IBM's Boca Raton, FL, laboratory provides the NetView/PC with the storage and operating cycles needed to execute the required input and output processing without greatly affecting the personal computer's performance.

When an application has data such as an alert for the host, it passes the data to the NetView/PC tasks via the application-programming interface. The data is then passed to the microcomputer interface, which causes an interrupt. The

*Part of IBM's Systems Network Architecture, LU 6.2 (also referred to as Advanced Program-to-Program Communications) defines a set of programming verbs that can be embedded in applications and called on when one device needs to access information located on another device. Users are employing LU 6.2 in a number of applications, including interconnecting equipment, providing messaging across various systems, and development of distributed systems.

The promise of LU 6.2 is that it allows peer-to-peer communications and the development of distributed applications. This can enable mainframe tasks to be offloaded to multiple smaller processors, improving host efficiency by freeing mainframe processing cycles and easing the burden of host communications associated with terminal support. LU 6.2 can also help users consolidate their networks by providing support for a wide range of devices. Users with multiple networks can reduce their network costs if they use LU 6.2 to bring these networks together.

Because LU 6.2 is capable of sending multiple streams of data simultaneously, personal computer users can have several processes executing in parallel. Many data-processing managers do not see much value in LU 6.2 because they have to rewrite their applications to take full advantage of its functionality.

LU/PU task then gets control and the data continues along to the coprocessor memory. Finally, the LU/PU task gives the data to the SDLC task to be sent to the host. Data from the host destined for the application uses the same route in reverse.

The hardware is a single-slot, full-length board that plugs into the PC. It uses a 16-bit 80186 microprocessor operating at 7.37 MHz and 512 kbytes of random-access memory. The card provides two serial input/output ports with full-duplex communications capability. Four electrical interfaces (RS232-C/V.24, RS422-A, 20-milliamp current loop, and V.35) allow the card to support most protocols. Communications between the PC and the card are through a shared storage-interface chip.

The multitasking operating system software is called the Real Time Control Program (RCP). It can control 253 concurrent tasks. Each task executes on any of 255 possible priorities.

The RCP provides functions that simplify the writing of tasks. This includes first-level interrupt handlers where the RCP receives the software and hardware interrupts first. It then passes them on to the task-interrupt handler.

A resource handler manages the various resources provided by the coprocessor such as hardware and software timers, queues, memory, communication ports, direct memory access channels, and interrupt vectors. A task can request temporary control of these resources.

Interface function routines allow application tasks running on the coprocessor to interrupt the PC in order to exchange data and control bytes.

The card is responsible for the NetView/PC host session. APPC, LU/PU runs on the coprocessor as a user task (FIG. 3-13). This task and the SDLC task are loaded and initialized when NetView is started. As part of their initialization, these tasks establish a session with the host.

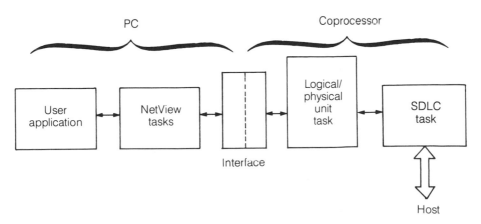

Fig. 3-13. Task allocation.

NETWORK-MANAGEMENT SUPPORT PACKAGES

A number of software packages extend IBM's network-management architecture to voice networks (FIG. 3-14). They allow users to produce billing reports for telephone calls, choose configurations for voice traffic, and troubleshoot Rolm Computerized Branch Exchanges (CBXs) from a central site.

An Alert Monitor allows the host to perform problem tracking and management. One Alert Monitor can serve ten CBXs and CBX II 9000 nodes in any combination. It can detect problem conditions, gather error and alarm information from the CBXs, and translate the information into different classes of alerts based on their severity. These alerts are then passed on to NetView/PC to be handled locally or to NetView for centralized problem management. The alert monitor can also use information obtained from a test package to report changing trunk conditions.

The Call-Detail Record (CDR) collector gathers information from up to ten CBXs. The program provides site collection, multisite polling, and host transfer of data that can be used for billing, capacity planning, and network analysis. Information can be stored and later forwarded to an IBM host for processing, using NetView file-transfer utilities via logical unit 6.2.

NetView can collect CDRs as well as error and alarm data from the CBX over an asynchronous connection. This information can be gathered and stored locally on microcomputer disk.

Several programs run in the NetView environment on the host:

- The Network Billing System is a System/370 application that provides users with an accounting of CBX costs and usage categorized by department, long-distance carrier, and equipment. The application package uses input CDRs that are generated by the CBX and then collected and forwarded to the host by NetView.
- The Traffic-Engineering Line-Optimization System allows users to analyze call-detail records received from the CDR Collector program. These

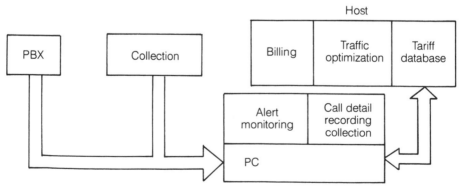

Fig. 3-14. Voice network management.

programs can be used to configure a site's telecommunication network for both voice and data. They also help identify problems as over/under capacity and isolated links.

A package that resides on the host is a tariff database that provides tariff information for long-distance carriers.

Switched network backup on the 5865 and 5866 modems allows the network operator to switch to backup dialed lines if a leased line fails. Up to four devices can share one remote Model 3 5865 or 5866 modem.

NetView has the ability to spot trouble along communications links using 5865, and 5868 modems with Link Problem Determination Aid-2 (LPPA-2) support for these modems. NetView contains a probable-cause algorithm that analyzes data collected from the modems forwarded to NetView by the Network Control Program (NCP).

The data reported by the modems provides configuration information to the probable-cause algorithm for analyzing problems on a link. This can be illustrated by supposing that a host is linked to several branch offices as shown below.

	HOST	
	APPLICATIONS	**NETVIEW**
	Virtual	
	Telecommunications	
	Access method	
Remote site	Operating system	
	Front-end processor	First link-segment Level
	Local mode	
	Remote modem	Second link-segment Level
Intermediate site	Local modem	
	Remote modems	
Branch offices	Remote devices	

One long-haul line, the first link-segment level, can serve up to four remote terminal sites via second-level link segments. The modems can report data to NetView from both link segments.

Commands available to the NetView operator include several that gather data about the modems and the line. The modem and line-status command retrieves modem and line data on variables such as line quality, random noise pulses, and the status of remote data-terminal equipment. This command also can be issued by NCP if a permanent link or station error is detected or a counter threshold is reached. Alert data is automatically sent to NetView.

A line-analysis command retrieves line parameters such as signal-to-noise ratio, phase jitter, and frequency shift.

The *transmit receive test* command causes a modem pair to exchange sequences of predefined bit patterns over a line. The test shows the number of blocks of data received with errors and the current transmission speed. Color encoding of bad, marginal, and good test results allows the operator to spot problem areas. The results of operator-initiated modem-and-line status commands are also passed through the probable-cause algorithm.

Other LPDA-2 commands control the modem's functions in the network. A modem control command changes the modem speed and controls a switched-network backup session. A modem configuration command defines and reconfigures the modem, setting the configuration to multipoint or point-to-point and the network role to primary or secondary.

An operator receiving a line-problem alert will have the cause identified (using the probable-cause algorithm), run the required testing of the line (using the line-analysis test), call the telephone carrier with data to report line problems (using data from the line-analysis display), and restore the network to operation at a backup speed until the repairs have been completed (using the modem-control command). These operations can be performed at the console by the NetView operator.

INSTALLATION AND OPERATION CHARACTERISTICS

In any complex network, an integral part of success is the ease with which packages can be installed and operated. NetView incorporates some enhancements in the areas of installation, customization, and operator training.

NetView uses an advanced installation process based on a prepackaged sample network. Simple procedures also simplify the user interface and reduce installation errors. The sample network includes Virtual Telecommunications Access Method (VTAM). Network Control Program (NCP), and NetView definition statements, along with tables, command lists (CLSTs), and job-control language (JCL). After installing NetView's functional code, the installer copies the entire set of network samples. The installer then verifies that NetView is functioning properly in the controlled environment of the pretested sample network.

Network installation has been reduced to a series of simplified steps. Structuring the task in this way makes many installation complexities transparent to the user.

The guide also leads the installer through the administration of installation. Administration tailors the product to support the unique configuration and requirements of the user.

In the administration phase, each step asks a question. These questions guide the user in modifying the samples.

The user's network is gradually built from the tested sample definitions. A structured approach can speed the installation, reduce the possibility of error, and focus resources on a more effective application.

Customization allows the users to tailor or supplement the product's functions to meet unique requirements. For a network-management product, these requirements might include accounting, collection of certain types of data for reports, and automation of processes. Users can customize NetView by writing their own exit routines, command processors, and subtasks. NetView also allow users to customize operator panels to increase operator efficiency.

The information that users can change includes help panels, hardware-monitor displays, and alert messages. A user can change the wording of a help panel to reflect a particular network configuration or change the color of text to highlight a unique operating procedure. Users can create new help panels for original applications. Users can also change the wording, color, and highlighting of the hardware-monitor displays and alert messages.

The operator also has the option of viewing any messages in the active log. NetView provides a full-screen browse capability through which files, as well as the log, can be examined interactively. Help panels can be looked over, as well as other libraries.

Log messages associated with important message indicators can be color-coded or highlighted on the log-browse panel. Browse panels use a scroll field and browse PF keys, as well as a string-find function.

An interactive help capability provides operator assistance in NetView. From the terminal, the operator can call up information on every command as well as components, recommended actions, and SNA and VTAM codes. The operator can also access information on specific screens and on panel customization. A step-by-step approach is used in a help-desk format.

To simplify network problem determination, complex procedures are reduced to simple steps. Diagnostic assistance also covers other products such as VTAM, the Network Performance Monitor, and other types of networks, such as 4700 Finance Communication Systems and 8100 Information Systems.

Users can customize the help-desk to match network configurations. The help-desk feature can aid new operators significantly in finding and resolving network problems.

The use of a command line on NetView panels and PF keys allows any NetView command or component to be invoked from any other component's screen. If an operator is displaying the session configuration screen in the session-monitor component, the status of the physical units can be obtained by entering a hardware-monitor command. NetView tends to legitimize the concept of network-management programs talking to foreign vendors' equipment.

About ten vendors have so far announced products supporting NetView/PC that provides a personal computer-based gateway for non-IBM vendors seeking entrance to NetView. Others seem to be developing interfaces for their products. But it is not clear whether these companies will devote the effort needed to develop host software to provide full network control.

NETVIEW EVALUATION

IBM will need to have a clearer strategy on how the mainframe will use the collected data. NetView should allow the user to send requests for tests such as modem diagnostics from NetView's console at the host to foreign vendor's net-management systems. This support is needed to make NetView much more powerful. NetView will grow in capability and most vendors of net-management systems who sell the SNA environments believe the marketing power behind NetView alone is enough to make NetView compatibility required. Users stand to gain not just from NetView itself, but from enhanced NetView capabilities sold by other vendors.

Several weaknesses of these products need to be addressed. One is the one-way communications problem. NetView and NetView/PC have a major shortcoming when they are used to provide network management for non-SNA products: communications is inherently one-way, from the remote devices supplying status information to the NetView in the host. There is no easy way for a NetView operator to issue control commands to non-SNA products in the network.

Another problem is the lack of Personal System/2 support for NetView/PC. Support is missing because NetView/PC requires a Real-Time Interface Co-Processor board that is not compatible with the Micro-Channel bus of most Personal System/2s.

IBM can transfer NetView/PC to the Operating System/2 environment, which is a multitasking operating system and more suitable than MS-DOS for the support of real-time communications.

IBM's network-management products also do not address customer requirements for usage-based SNA network chargebacks. This becomes increasingly important as companies integrate their data communications into a single corporate utility SNA network. These companies need a way to allocate network costs based upon the amount of traffic being transmitted. NLDM, which is integrated into NetView, provides only 3270 traffic counts. The Network Performance Monitor (NPM), measures traffic volumes separately for each 37XX/Network Control Program (NCP) node. It consumes 10 percent of the throughput of customer front-end processors when simultaneously monitoring traffic volumes on all circuits.

It also requires custom applications in the host that can process and piece together data collected by NPM from multiple NCP nodes. That can be a major undertaking. The ability to collect traffic volumes for chargeback purposes is standard in X.25 packet network products.

In spite of these weaknesses, NetView and NetView/PC might become de facto standards because international standards for network management will appear later. A de facto standard could develop because open systems interconnect (OSI) standards for network management will appear more than two years later. The availability of products implementing these standards is even further off. The OSI Management Framework and Common-Management-Information-Service protocols, which are international standards, are due in the summer of 1988, and the five specific management-information services and protocols necessary to implement products are expected to become standards in late 1989.

By that time, products implementing IBM's network-management architecture will be widely used, particularly by SNA network customers. Thus, NetView/PC is positioned to lead the industry in network management because of the current lack of OSI network-management standards.

HUMAN FACTORS

As computer networks evolve to meet the needs of those that create them, they tend to affect their user's social structures. A firm's experience with a new network is the meeting of essentially two different sets of social processes.

Network management requires an understanding of the social processes so they can select networks that are harmonious with those processes. Networks exhibit the social structures that surround and influence them.

Examples include the differences between IBM's Systems Network Architecture (SNA), a rigid hierarchy, and Digital Equipment Corporation's (DEC's) Decnet, a peer-to-peer networking scheme. The architecture of a network tends to match the organizational philosophy of the company, whether that is centralized, decentralized, or distributed.

This is also shown in packet-switched data networks. Arpanet was built to support military needs. The architecture permits new nodes to be added and removed easily, providing a flexible topology. This made the network popular with researchers, generating many implementations on college campuses. Military and classified users have thus broken away and formed their own subnetwork. Telenet is a commercial spin-off of Arpanet, built to conform to standards such as X.25. Tymnet grew as a more ad hoc solution to time-sharing access needs. The network was built using proprietary internal protocols. New needs were met with tools developed for the network. Telenet and Tymnet grew more alike with time, but some of their different origins remain.

Computer architectures tend to reinforce the structure of the user organization. SNA was originally favored by highly centralized companies. Decnet is preferred by engineering firms. This grew out of an environment in which all of the computers had an equal chance to express themselves.

There has been a decline of centralization, evidenced by the trend toward fewer mainframes and more mini and microcomputers. The times are prompting the move away from centralization. Top-down communication can be very effective in a military, religious, political, corporate, or other conventionally structured organization. However, when changes in organizational culture, climate, structure, or process take place, top-down communication paths often can be inadequate.

One way of achieving this is to build, or adapt, existing hierarchical structures for use as overlays. An example of this is the hierarchical supervisory overlay network, which is used in packet networks with a distributed-mesh topology. In a similar way, IBM's NetView tools can serve as an overlay for control networks of non-SNA equipment.

PROTOCOL DIALOGUES

Communication protocols in network architectures reflect human communications at each layer. Network protocols are like human dialogues. When two users or applications use a layered architecture, their interactions are conveyed by messages passed between and within the layers.

This is true of computer use in general. Clicking a mouse somewhere on a screen activates code that sends a notice to a subroutine along with the parameters stating the cursor location.

The interface between the calling module and the subroutine has been agreed upon by the programmers. This type of relation is even clearer in communications. In the physical layer, for example, names of RS-232-C leads and the handshaking sequences that use them suggest people using the telephone:

- Ring indicator—an incoming phone call
- Data Terminal Ready—the called party picks up
- Carrier detect—ability to hear each other
- Link-level protocols support more complex dialogue

Byte-level messages such as ENQ, ACK, and NAK evolved into bit-level equivalents with the same meaning. The meanings and proper use of these messages must have been agreed upon beforehand.

More complex multidrop link disciplines may also reflect attitudes about communications. Hierarchies operate in a military fashion, a polling protocol resembles a sergeant questioning a row of privates.

Token-passing protocols resemble a game where messages are passed from person to person. Collision protocols favored are more crowded with everyone trying to be heard. They are more chaotic but potentially more democratic.

Intermediate-layer protocols, such as those in the International Organization for Standardization's Open Systems Interconnection model, deal with across the network through the entity at the layer below.

The dialogue can be as follows:

- Request—A asks its underlayer counterpart to convey a request to B.
- Indication—B's underlayer counterpart informs B of the request.
- Response—B gives its underlayer counterpart a response for delivery to A.
- Confirmation—A's underlayer counterpart relays B's answer to A.

Such sequences can be used for requesting network-layer calls, resetting transport-layer synchronization points and negotiating session-layer services.

The application layer, like the physical and link layers, tends to be busy. Individuals and organizations usually have more than can be handled by one protocol. A number of protocol standards at this level is often needed.

PERSON-TO-PERSON ISSUES

Many protocols are designed for computer-derived uses such as file transfer and virtual terminals. Some will be used to support interpersonal messaging. Standards that are intended to support actual users should include an underlying view of people in organizations.

Person-to-person standards of communications should allow users to interact in a natural way. Engineers design networking software, but the ultimate users might work and think differently.

The message oriented systems need to contain:

- Simple user conventions
- Automatic pickup and delivery of messages
- Error-checking with recovery and redialing
- Network administration and accounting
- Background communications services

An example of the use of these concepts can be found in a development tool called the Message-Handling Service (MHS), which Novell offers for such applications as distributed databases and accounting programs. Novell's MHS should not be confused with the Message-Handling System [also called MHS] part of the X.400 international standard. MHS is consistent with X.400's MHS and can produce compatible packets.

The Novell MHS package runs on Novell's Netware and several other network operating systems. These include 3 Com's Ethershare, Banyan's Vines, IBM's PC LAN, and any DOS 3.2 compatible LAN.

MHS can support as many as ten thousand users. A workstation on a local-area network would run software that can access records of the user's conversation and a calendar, as well as DOS (Disk Operating System) files, on the file server using a configuration such as the one in FIG. 3-15.

Novell's Message-Handling Service lets users send local or remote messages by specifying the recipient's name.

As a user posts a message, it is formed by the sealer and shipped to the file server. Messages are sorted, addressed, routed, bundled, and queued by the server, which can be a dedicated microcomputer. A module called the *router* or *address manager* keeps a directory of users and how to deliver messages to them.

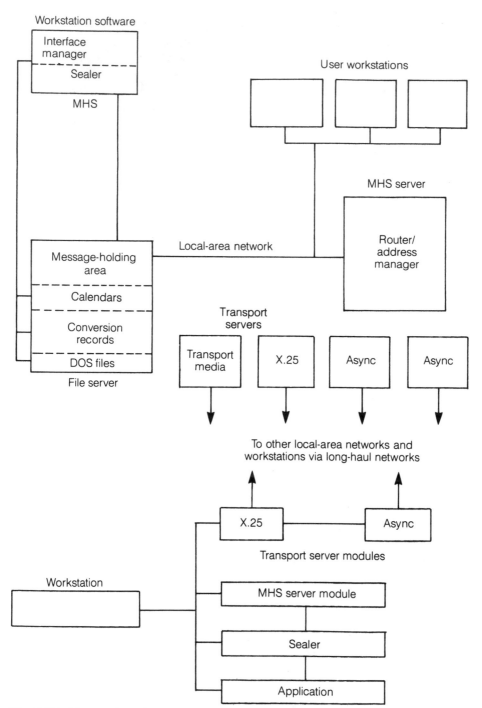

Fig. 3-15. Message-handling-service configuration.

Messages destined for other workstations on the LAN are delivered immediately. Outbound messages are sent using a transport server for the message pickup and delivery. This is done either periodically, at certain clock times, or after a given number have accumulated. After delivering its own messages, the originating server checks for mail waiting to be picked up. Transport servers can reside on the same machine as the communications server or on dedicated hardware. All modules can reside on an independent or solo workstation.

A module called a *converter* is needed to converse with IBM's Profs, DEC's All-in-One, MCI Mail, Telenet's Telemail, or others.

A semistructured approach will let users compose messages and select, sort, and prioritize them.

User-specified rules include the following examples:

- If an action request has a deadline, it is moved to an urgent folder.
- If an information request concerns a particular topic, it is moved to an "urgent file."
- If the subject is a meeting, it is moved to a meetings folder.

Messages also can trigger automatic responses. A meeting proposal might have a rule that automatically constructs a meeting acceptance message, and upon receiving the recipient's approval, forwards it to the sender.

When the rules concerning operations are carefully constructed, automatic messaging becomes a part of a distributed artificial intelligence.

Users can build their own filters that screen information. Another mechanism lets users find messages that they are interested in from a pool of public information.

One view taken about the use of computer networks to support human ones is that people make their speech act explicit, but computers can only amplify and coordinate human commitments. The use of AI gives computers more responsibility in the content of people's work.

For example, electronic mail can support not only passive messages, but act as intelligent messengers that perform tasks in the network. This might include setting up a meeting checking the workstation of each person on a meeting list, or checking a calendar to see if that person is available.

It might try to reschedule the meeting by revisiting previous workstations. If it had no rules concerning what to do, it might return to the sender and request some.

It might also help the user process messages and suggest likely responses to certain messages and offers to take actions based on them. For example, after presenting a meeting announcement message, it might ask if it should add the meeting to the user's calendar.

Distributed artificial intelligence (DAI) gives computers a more active role in creating, manipulating, and conveying meaning.

DISTRIBUTED ARTIFICIAL INTELLIGENCE

Networks will include more and more of these entities that mirror the world of professional interaction. Networks increasingly will support this form of near-human dialogue that might become future network protocols.

Expert systems will also be called upon to solve a variety of localized problems, such as those concerned with individual callers, network components, failures, and demands. They will take advantage of the distributed approach.

The problem-solving entities must work together toward a globally optimal solution without duplicating each other's effort. Expert systems should be able to distinguish between those situations that require them to work together and those that do not.

The advantages of distributed AI are similar to those of distributed processing in general:

- The computing power of small, inexpensive processors can be harnessed.
- Local resources can be applied to local problems.
- Reliability is improved by avoiding a single centralized unit that could fail and take the network down.

The techniques of distributed expert systems can be applied to network-management problems to automate network operations. Knowledge-based systems can observe network activity and modify the network to handle traffic. Autonomous, self-healing networks are a logical end product.

Network management will require a number of expert systems to handle such areas as:

- Tariffs and line quality. To help users choose between lower cost and high quality, an expert system could monitor requests for service and fulfill them based on the resources.
- Diagnostics. Each element in a network, such as a trunk or a multiplexer, could have its own expert system for testing.
- Maintenance. A maintenance expert system would examine patterns of failure and suggest maintenance rather than repair action. AT&T's Automated Cable Expertise (ACE) is an example of an expert maintenance system.

These systems could also solve practical networking problems, such as shadowing, load balancing, and performance tuning and monitoring. Disputes between the expert systems would have to be resolved by another expert system or by a human intervention.

Most expert systems are based on the knowledge of real experts, which is then customized. Although many operations are common to all networks, local ways of doing things have to be part of the AI network. Parameters have to be set from information that the user provides by filling out forms and menus in the installation process. This process itself is handled by an expert system.

This decision is not always finalized when the systems are initially set up. The organizational design can be monitored and updated over time.

One aim of DAI is to create problem-solving entities that can organize themselves in optimal ways based on their understanding of the problem as groups of people come together in different ways.

To illustrate how organization can relate to tasks, consider the organizational designs of fault analysis, which lends itself to a hierarchical organization (FIG. 3-16).

Fault analysis is accomplished with a number of expert systems implemented as local fault analyzers at network nodes. Each expert system receives error messages from local facilities and uses these to diagnose local failures. Messages relating to regional failures are passed to regional expert systems, which in turn pass network-wide problem messages to a top-level analyzer. This hierarchical arrangement of intelligence is still a distributed organization.

Virtual private-network management is more suitable to expert systems that function peer to peer (FIG. 3-17).

Software-defined networks can have their own virtual private subnetworks. These could be managed by separate expert systems. These expert systems might function as peer-level entities. Local changes would be handled by each

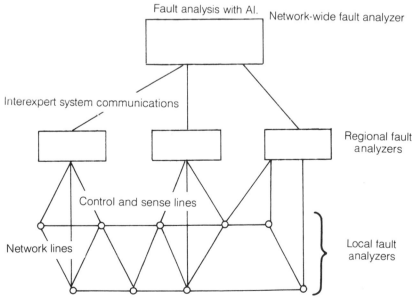

Fig. 3-16. Fault analysis with AI.

Virtual network management with AI.

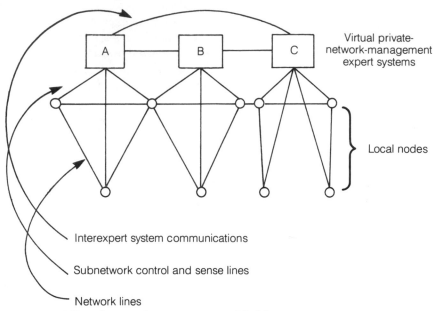

Fig. 3-17. Virtual network management with AI.

expert system independently. Before making changes affecting other sub-networks, an expert system would confer with its peers.

Suppose subnetwork A had more traffic than it could handle, its management expert system would then ask its peer in subnetwork B for unused resources.

4

Network Measurements

The concepts of network testing and measurement/monitoring capability that should be implemented to maintain a digital communications system are often critical to its success. The combination matrix shown in FIG. 4-1 suggests the relationships between network diagnostic methods.

AN INTRODUCTION TO NETWORK TESTING

Surveillance and testing are both methods to access information. Surveillance is unobtrusive and refers to the use of equipment on circuits without affecting their performance or user service.

Surveillance can be continuous or on a scan/sample basis occurring only when a result deviates from the norm. In contrast, testing is more disruptive; a circuit must often be taken out of service and reconfigured to apply a reference signal and a detector.

Both monitoring and measurement allow the use of the acquired information once surveillance or testing access has been gained. Monitoring can be used to compare discrete events that have been recognized as unusual, such as failures or signals crossing preset thresholds. Measurement refers to the selective collection over time and the data is often analyzed later.

Systems that allow monitoring and measurement are typically two-way systems. They send data for analysis and then accept signals to make changes in the operation of the network, such as switching in backup circuits or turning off audio alarms.

Terms such as test and monitor appear unambiguous and straightforward, but often what is needed is standardization of the terms with which the network diagnostic procedures are defined.

It is desirable to avoid large numbers of users out of service simultaneously while problems are isolated and fixed. Technicians need to know where incipient

	Monitor	Measurements
Surveillance	Detect Alarms	Evaluate performance
Testing	Fault location	Troubleshooting

Fig. 4-1. Network diagnostics.

service problems are before they become apparent to users. Technicians must also make the trouble conditions on high-capacity systems known to the service provider's personnel to speed restoration. Telemetry hardware can meet some of these needs. For example, bit error rates on T-1 facilities, errored seconds, and bit error rates directly in the hardware all can be measured. The major issue is how to combine this information to provide an orderly maintenance plan for the network.

Measurement with testing access is familiar and often the most expensive combination. It is typical of analog service and low-speed digital service (64 kbps and below). Maintenance is performed with a known signal injected by the test equipment. The signal is usually a pseudorandom bit sequence that the receiving test equipment can duplicate for purposes of detecting bit errors. Monitoring with surveillance access is typical of most current alarm surveillance and control systems as well as performance-monitoring systems.

Monitoring with testing access is a more uncommon maintenance approach. It can require removing a circuit from service and applying a known signal.

A measurement with loopback devices deployed at various points along the circuit usually is preferred over circuit removal. It is used routinely to check circuits.

An alternative to full-blown measurement with testing access is measurement with surveillance access. This has not been practical in analog service applications because it would involve measuring analog impairments on a live circuit in the presence of strong speech or voiceband data signals. This is only now coming within reach with digital-signal-processing technology.

A digital circuit can be tested with a line-performance monitor. A performance monitor, perhaps a scanning device that can monitor T-1 lines, can monitor a line that would need testing because of a trouble report or surveillance-monitoring result. This allows longer performance measurement without service interruption. The user's data can also be used for T-1 testing rather than a test pattern.

AT&T's Bell Labs found that lines selected on the basis of high in-service error rates were frequently error-free when they were stimulated with a pseudorandom test pattern. When the lines were put back into service, the errors returned. This suggests a bit-pattern dependence of the errors.

In digital computer maintenance, bugs are often found in the hardware or computer logic using a part of the user program in a test driver. The test driver is run many times to isolate the problem. An analogous approach might be needed for communication's circuits.

At line rates of DS3 and greater, surveillance can take two forms. Line monitoring using the error detection features of the DS3 framing scheme can be used. The use of an intelligent network element (multiplexer or digital cross-connect) in the path allows a comparison of the performance registers in that element over periods.

Surveillance measurement is an excellent alternative for test measurement at DS1 rates and above. Surveillance can often eliminate the need for testing by associating the proposed measurement with a known failure.

When a user initially orders service and the lines are installed, testing is necessary to assure that the service is at the level required. First, a test measurement is made. After a period of time the circuit might degrade, and an alarm can be set off by the network management system to indicate that service is still at this level, but there is no safety margin. This is the essence of surveillance monitoring.

The network-management system can also direct the monitoring equipment to listen to the circuit where the alarm originated. This surveillance measurement could indicate a problem, and the user would contact the service provider for repairs.

A surveillance measurement can then be made to see if the circuit is again within the tolerance limits. A test measurement and the associated user outage could be avoided unless the user desired a standard test to verify service.

The application of testing and surveillance will vary with the user organization. The ability to schedule repairs on a preventive basis through the user of an integrated measure/monitor capability will eliminate expensive routine testing and endless redundant testing.

NETWORK PERFORMANCE ANALYSIS

In general, the analysis of network performance should proceed from the lowest layer of the OSI model to the highest layer. The parameters at the lower layers of the network provide a basis for understanding the higher layers. A first step in analyzing a network is to study the network's overall health. This can be done by measuring the network's current utilization (continuous and peak), traffic variation over the time, and the types of data-link errors that can occur.

The utilization on an IEEE 802.3/EtherNet network can be plotted as the time the network is busy versus the time the network could be moving data within the rules of the access method. This can be plotted as a percentage of utilization in frames/sec. or bits/sec. Frames can be a reasonable measure of arrival times for a node, but it is not a good measure of network utilization due to the wide variations in frame lengths. Other measures can be used.

The carrier-sense multiple access with collision detection (CSMA/CD) method used in 802.3/EtherNet-based networks has no explicit overhead frames to support the media access method. The total traffic measurement provides a better indication of work being done in the network. The utilization average and the instantaneous peaks can be used to estimate the network usage. For utilization of the network the percentage utilization or aggregate bit rate can be used. For a node's ability to process incoming data, the rate at which frames arrive can be important.

NETWORK UTILIZATION FACTORS

Utilization varies dramatically with time and is influenced by user work habits. Activity on a network corresponds to the work habits of users working on specific network nodes.

Trends include peaks at mid-morning and mid-afternoon. Other factors often contribute to peak utilization levels. Backups of large databases, loading, and unloading across the network for processing can occur during evening hours, causing additional peaks in utilization. FIGURE 4-2 shows these characteristics.

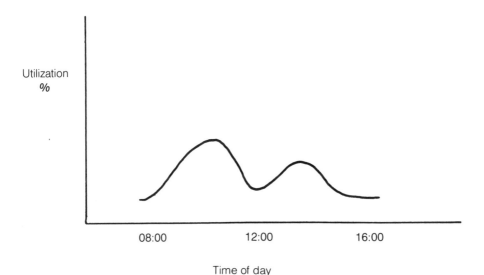

Fig. 4-2. Network utilization.

A network often consists of a number of workstations. File servers provide the access to shared storage, laser printers, and plotters. Terminal-to-host virtual services usually exist all across the network.

In a typical 24-hour period, there might be little activity in the early hours of the day. Starting at about 7:00 AM, activity begins to build to a peak around 10:00 to 11:00 AM. Utilization of the network then builds to another peak at about 2:00 to 3:00 PM, and then tapers off until the evening hours. During the evening some major file transfers might take place between hosts on the network.

Depending on the network, utilization cycles exhibiting daily, weekly, or monthly characteristics can occur. Administration-induced peaks can occur on financial cycles such as the end of the month, quarter, or year. Engineering-induced peaks can occur with project deadlines and midpoints. Manufacturing-induced peaks may occur with processing starts and stops. Depending on the internetworking scheme and geographical time zone, change can also have an impact on the profile.

Network performance is often a function of error conditions on the network. In IEEE 802.3/EtherNet networks, basic errors can occur at Levels 1 and 2.

The *frame-check sequence* or *cyclic redundancy check* verifies that the bits in the frame were transferred correctly. A misaligned error occurs when a frame is received in which the total number of bits is not divisible by eight.

Such errors can be caused by several conditions. Illegal-size frame errors can be of two types. Frames of less than the 64-byte minimum are sometimes called *runts*. They are usually caused by collisions on networks over large distances. The problem occurs on large networks when one node begins transmission and another node physically distant from the first begins transmission and collides with the first. Runts occur because of the propagation delay on the network which allows the first node to transmit enough of its frame onto the network to be recognized as a partial packet.

Frames longer than a maximum of 1,514 bytes are sometimes called *jabbers*. These are usually caused by hardware failures. Runts represent future potential problems that require further investigation.

Collision on IEEE 802.3/EtherNet networks are a part of the access technique. In light traffic situations (under 5 percent) collisions occur as a function of the randomness of the data.

Network-loading information has several important uses. It facilitates the pinpointing of potential problems, and if it is coupled with traffic source and flow data, it provides the network management information needed to make reconfiguration decisions.

When analyzing traffic sources, the important factors are source-destination address pairs, multicasts (packets with addresses that are destined for a group of devices), and broadcast addresses. Hosts and file servers tend to be the concentration points for such traffic.

When analyzing specific nodes, search for those with high levels of incoming and outgoing traffic. The activity of a particular node can indicate that further analysis is required. A plot of source addresses on the x-axis and destination addresses on the y-axis can be set up as shown in FIG. 4-3.

Each frame sent from a given source gives a number at the intersection. The number indicates the frequency of traffic.

OSI level-2 address pairs can provide a basis for reconfiguration analysis. The important factors are:

- Data rates
- Error rates
- Frame size activity

Level-2 activity can change quickly depending on user activity, but often only a few connections generate most of the traffic. These usually will be the site of problems if they should exist.

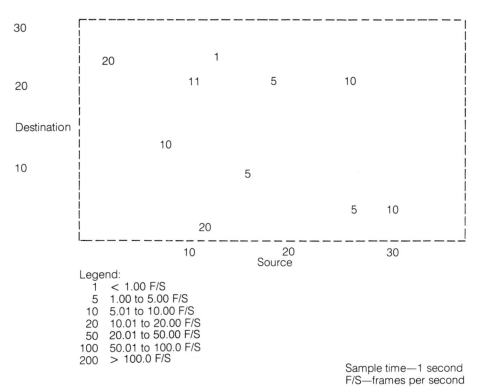

Fig. 4-3. Communications matrix source vs. destination.

USING BRIDGES

Bridges can be useful tools in configuration management. In the OSI model, a bridge provides a conversion point between two networks at the link level. It is generally not sensitive to upper-layer protocols. The information field at the link is just a group of bytes.

Gateways allow true conversion between two protocol environments and can be used on any layer. Bridges and connection-analysis information allow networks to be configured to funnel traffic to specific segments in order to minimize internetwork traffic.

Bridges used to link two networks are programmed as to which nodes are on each side of it. The function of the bridge is to forward traffic from one network to the other according to the address. Bridges allow the network to support higher aggregate loads. They can also be used to isolate physical problems between segments. They must have adequate capacity or they could create bottlenecks.

Other traffic characteristics include the size of the frames, their sources, and the burstiness of traffic. Network services such as virtual terminal, file transfer, internetwork connections, and higher-level transport protocols such as Xerox Network Systems (XNS) can all affect the frame size.

Different network services generate different frame sizes. Terminal servers tend to use small, minimum-length packets that allow the network to be responsive but waste capacity. File-transfer applications tend to use packets of maximum length and so are more efficient, but they can block network access.

Asynchronous terminal servers, which interface terminals, tend to transmit traffic to an upstream host in small packets. In the downstream transmission from the host to the terminal, the packets tend to be larger.

FRAME SIZE

The average frame size varies throughout the day. During the prime time of the day, the average frame size remains relatively small. During the evening hours the average frame size can increase dramatically because of the bulk batch transfers that are configured to maximize the frame size.

Internetwork traffic can also influence frame size. If a network supports only small frame sizes, one message can be split into several small frames. An example could be a gateway to an X.25 network that interfaces to an 802.3/ EtherNet LAN. If the X.25 network is configured to support packets of less than 1500 bytes, the maximum-length packets coming from the LAN will have to be broken into multiple X.25 packets prior to transmission. If those smaller packets return to a LAN through another gateway, they will increase the probability of network delay.

As the size or traffic increases, delay measurements become more important. Two measures of network delay are the time needed to access the network and the propagation delay between nodes. Network-access delay is an important part in the transmission delay caused by the physical network.

The components that affect response time include communication processing time on each end node, host processing time based on the transaction, and propagation time of the total communication network.

LOOPBACK COMMANDS

Many protocol and network standards use specified commands and responses for testing protocols and software. In addition to loopback testing procedures for upper-layer protocols, several local-area network standards include level 2 loopback procedures. Two examples of such test facilities are the EtherNet Configuration Testing Protocol (CTP) and the IEEE (Institute of Electrical and Electronic Engineers) 802.2 Test Command and Response.

The EtherNet CTP allows single or multiple-hop loopbacks. With this facility, a frame can be sent to a node, and then either returned or forwarded to a third node, returning finally to the originating node.

The IEEE 802.2 test facility involves sending a test command frame to a node. If the command is the proper format, the node will return a *test response*, which contains an information field identical to the original test command. These facilities can be used to test for the proper functioning of the level-1 and level-2 network interface.

PROPAGATION DELAY

Propagation-delay measurements show how performance varies with network usage and give a good indication of potential performance and capacity problems.

To measure propagation delay, it is necessary to compare the time the message is successfully placed on the network to the time the response message is successfully received. The remote node-processing time is subtracted, and the remaining time represents the round-trip propagation delay for the network. Measuring this variation over time under a variety of network loads will indicate the network's contribution to transaction response time.

Propagation-delay measurements are particularly helpful at gateways and bridges. By taking the measurements over long periods of time, it is possible to determine how performance varies with network usage and to flag potential performance and capacity problems. Suppose a plot is made of the results of sending a periodic loopback message to a node on a remote network connected to a LAN via a bridge. The plot displays the total time from successful transmission to receipt of the response. Subtracting the processing time on the remote node gives the total round-trip response time of the network.

To effectively measure network characteristics, it is necessary to establish a baseline of information about the network. The baseline consists of the network performance information. Baseline characteristics can include ongoing utilization, peaks, and error rates.

Another network traffic parameter that can influence errors and waste network capacity is the burstiness of the traffic. When a network node receives a large number of packet bursts, it is possible for that node to miss some packets. This occurs when the node or its interface is not able to process the incoming packets fast enough.

These errors are masked from the user by the upper-layer protocols. Protocols retransmit the data one or more times until it is properly received. Any retransmission wastes network capacity and affects node-response time.

The network baseline can serve many purposes. It can provide a basic reference of data to explain the general network characteristics. It can also be used as a reference in troubleshooting performance problems, and it can be used to estimate the effects of future growth.

Modeling tools and techniques are important tools for analyzing anticipated growth.

PROTOCOL ANALYZERS

An aid to modeling is to use a LAN protocol analyzer. This type of instrument can be used to provide an independent, controlled network load. The analyzer can be used to observe the effects on network performance parameters. Simulating traffic and measuring its effects can provide information on the impact of growth and changes.

When a simulated traffic load is used, it is important that it have the same characteristics including the distribution of frame size and burst rate. Adding traffic with representative characteristics makes it possible to predict performance error rates and collisions.

Modeling the traffic target for or addressed to specific devices and measuring their response is useful for identifying the limitations of such devices as bridges and gateways.

Many firms that have been using EtherNet local-area networks are finding that managing them can be challenging. There are a number of factors that one needs to know:

- What the network loads are at any particular time
- What channels are most active
- What types of errors are occurring
- What will happen when more terminals are added

Products like Excelan's LAN analyzer and HP's HP 4971s provide network performance statistics and can be used to run tests on individual stations and review the results. They can analyze high-level protocols such as Transmission Control Protocol/Internet Protocol and Xerox Network Systems.

They cannot simulate a terminal to perform extensive testing, nor is it possible to remove faulty products from the network with this type of analyzer.

Most products of this type interface with the local-area network backbone and provide menus that permit the operator to establish test parameters, take measurements, and evaluate the results. Users can receive help screens and on-line portions of the user's manual.

The different packaging approaches allows buyers to choose from several alternatives. One is a controller board that fits into an expansion slot in an IBM Personal Computer, XT, AT, or compatible with at least 512 kbytes of random-access memory.

Packaged systems can consist of an AT type of personal computer with 640 kbytes of RAM, a 20 megabyte hard disk, and the installed controller board. Local-network-control software is also available separately on diskettes.

Units are available with a communications facility and statistical package. Some work only as an analyzer, while the others can function as an EtherNet node or stand-alone personal computer when not being used for network analysis.

Most analyzers provide the same types of tests and furnish the same levels of information. Operators can name the channels or individual terminals to be tested and the types of errors to be sensed and collected. Users also can specify their own specific bit-pattern triggers.

The user can set limits on the size of the packets to be captured to eliminate clogging the capture buffer with small packets caused by collisions. There can be a preset limit on packet size of, for example, less than 512 bits. The rationale for this limit occurs in heavily loaded networks. When short packets are received with a minimum interframe spacing, many local network interfaces can miss the packets. In many cases, the user would prefer to have the option to specify this parameter. Typically, dual-ported, 1 megabyte circular capture buffers are used to hold the captured packets. Users can limit the volume of data stored by specifying the channel filters that are matched against incoming packets.

Most units chain buffers for handling captured packets. Buffer boundaries can range from 256 to 480 bytes.

Users can specify the start and stop times for tests and the number of packets to be stored in the capture buffer. They can also have the option of storing and recalling either entire packets or portions called *slices* on hard disk or floppy.

Many products allow data to be displayed simultaneously in hexadecimal and ASCII. The displayed information can indicate the source and destination address for each packet, the length of the packets, the type of errors detected, and the time the packet was received. These units allow the operator to scroll through the entire buffer or go to selected source and destination addresses within the buffer or to the packets on disk.

NETWORK-MANAGEMENT FUNCTIONS

The network-management capabilities of most testers are restricted to measuring the traffic loading on each channel. Most allow users to inject packets into the network data stream to test how well the system reacts to increased traffic.

These analyzers allow users to view traffic-level loading in real time. The screen display can show the following data:

- Channel loading as a bar graph
- The percentage of errors occurring by error type
- The number of packets transmitted with and without errors
- A count of individual error types occurring on each channel
- The number of complete packets received

Also provided is a display that can show the source and destination addresses of user-selected channels and the volume of packets sent by one address to another. This allows the user to assess the interchannel traffic distribution in real time.

Bar graphs showing burst-traffic periods measured according to interframe-spacing time allow network planners to view the traffic patterns in real time. They can also check network response time to determine message-transmission time. The display can show absolute time, acquisition time in microseconds, the percentage of deferred requests, collisions, and percentage of aborted attempts. Bar charts can show network loading during specific periods. The display can show other useful information such as the average packet size gathered over a specific test period and present data as a percentage of bits in each packet. If 95 percent of the packets had 64 bytes, this is displayed along with the peak traffic rates. Performance statistics include the number and types of errors occurring during the measurement period.

In addition to analyzing on-premise local networks, many products can be used to control measurements at remote locations. Software packages such as Meridian Technology's Carbon Copy can be used to control the remote site.

Most products do a reasonable job of measuring traffic, provided the volume stays below 40 percent of network capacity. Above this limit they tend to lose packets.

Packets can be lost in special applications such as setting traps for specific bit patterns. Because setting data traps is a common fault-analysis application, losing packets can be serious.

In summary, while most units provide approximately equal facilities for monitoring and measuring local-network performance, some screens incorporate more related performance statistics. Customers have the option of using lower cost or pre-owned hardware, and some units can be employed as a network node or other applications when not being used for analysis.

When choosing an analyzer, look for one that is user-friendly and furnish all the needed network tools. Packaging, portability, and price are other variables that must be considered.

FIGURES 4-4 to 4-10 represent screen views that illustrate some features of a typical protocol analyzer. They are intended to convey the flavor of a live analysis session.

NETWORK HARDWARE TEST REQUIREMENTS

During the process of installation, operation, and maintenance of the network, the user must be able to test, monitor, and control the system. This test capability should be sufficient to allow the verification of the system specifications during implementation. After the system has been implemented, the hardware components can be monitored on a periodic basis and controlled to ensure that the operational specifications continue to be met.

Hardware is subject to various sources of degradation such as aging and environmental effects. Real circuits contain real components and they will depart from the hypothetical ideal circuits used in the design methodology because of differing conductor length, media characteristics, and other factors. These differences become important during the last stages of troubleshooting and performance evaluation. This section discusses techniques for testing digital networks for parameters such as bit error rate, jitter, bit-count integrity, and equipment reliability.

Hardware test methods generally require rigid performance or standards that can be verified by testing, either by the vendor or the user in the factory or in an operational environment. To facilitate the testing process, the specifications and standards usually should include descriptions of the tests to be used as well as the recommended test equipment, configurations, and detailed procedures.

BIT ERROR RATE MEASUREMENT

Many performance monitors that allow system and equipment margins to be monitored continuously use an estimation of the bit error rate. The *bit error rate* in a digital transmission system can be expressed as the ratio of the number of bits received incorrectly to the total number of bits transmitted. The error parameters used in measuring system performance are often the same as that used in the system design to allocate performance.

The measurements can be made under out-of-service and in-service conditions. In the out-of-service case, the operational traffic is replaced by a known test stimulus or pattern. This is often a pseudo-random binary sequence (PRBS) that is used to simulate the traffic.

Summary Delta

M		Delta	Source	Destination	Protocol
	1	0.089	SNA Gateway	SNA Gateway	NET Name SNAGATE17 recognized
	2	0.100	NETBIOS	SNA Gateway	NET Find name GUESTC
	3	0.706	NETBIOS	SNA Gateway	NET Find name MIKELEOC
	4	0.000	PC Server #2	GEC	LLC C D=F0 S=F0 RR NR=42 P
	5	0.022	GEC	PC Server #2	LLC R D=F0 S=F0 RR NR=42 F
	6	0.064	NETBIOS	SNA Gateway	NET Find name JIMBERRC
	7	0.000	PC Server #1	Hi Mom	LLC C D=F0 S=F0 RR NR=117 P
	8	0.104	Hi Mom	PC Server #1	LLC R D=F0 S=F0 RR NR 123 F
	9	0.000	NETBIOS	SNA Gateway	NET Find name SUZANNEC
	10	0.105	PC Server #2	HJS	LLC C D=F0 S=F0 RR NR•31 P
	11	0.000	HJS	PC Server #2	LLC R D=F0 S=F0 RR NR•113 F
	12	0.124	NETBIOS	SNA Gateway	NET Find name BUCAYC
	13	0.099	NETBIOS	SNA Gateway	NET Find name SNAGATE17
	14	0.010	SNA Gateway	SNA Gateway	NET Name SNAGATE17 recognized
	15	0.090	NETBIOS	SNA Gateway	NET Find name GUESTC
	16	0.052	PC Server #1	PC Server #1	LLC C D=F0 S=F0 RR NR•69 P
	17	0.001	PC Server #1	PC Server #1	LLC R D=F0 S=F0 RR NR•69 F

1 Help	2 Set mark	5 Menus	6 Display options	7 Prev frame	8 Next frame	10 New capture

A Summary view: each line represents one frame, shows frame number, time since the previous frame (to a millisecond), DeSTination and SouRCe address, highest level protocol, plus other details.

Figs. 4-4 through 4-10. Screen views of typical protocol analyzer. (Screen format © Network General Corporation.)

```
CAPTURING                    Number of frames from the station              00:00:08
 Eagle One      118      118   IBM Portable This Sniffer   1              Broadcast
 IBM Portable     1                Broadcast   Eagle One    1             Broadcast
 42608C187062     1                Broadcast

240 frames seen.                       9 Kbytes,        240 frames accepted.   98% buffer used.
      3             10                      30                   100      300         1000

                                  Frames per second

1             3 Data        4 Clear        5           6 Captur                 10 New
Help          display        screen       Menus          options                  capture
```

Capturing frames from a network lasts just a few seconds until the end of file is reached.

The data fields shown at the top of the screen provide activity information on each active node of the network during the capture period.

The thermometer display shown indicates overall network activity.

Fig. 4-5. (Screen format © Network General Corporation.)

Traffic Generator	Y	Station address	Y
Capture filters		Protocol	Y
Trigger	Y	Pattern match	Y
Capture	Y		
Display	Y		
Files		From <any station>	Y
Exit		‾o <any station>	Y
		± Reverse direction	
		± Non-broadcasts	
		± Broadcasts	
1	3 Data		10 New
Help	display		capture

The From and To station selections of the menu permit you to specify the particular station(s) whose traffic will be captured.

Fig. 4-6. (Screen format © Network General Corporation.)

SUMMARY	Delta			
1		SNA Gateway	SNA Gateway	NET Name SNAGATE17 recognized
2	0.089	NETBIOS	SNA Gateway	NET Find Name GUESTC
3	0.100	NETBIOS	SNA Gateway	NET Find name MIKELEOC
4	0.076	PC Server #2	GEC	LLC C D = F0 S = F0 RR NR = 42 P
5	0.000	GEC	PC Server #2	LLC R D = F0 S = F0 RR NR = 42 F
6	0.022	NETBIOS	SNA Gateway	Net Find name JIMBERRC
7	0.064	PC Server #1	Hi Mom	LLC C D = F0 S = F0 RR NR = 117 P
8	0.000	Hi Mom	PC Server #1	LLC R D = F0 S = F0 RR NR = 123 F
9	0.104	NETBIOS	SNA Gateway	NET Find name SUZANNEC

DETAIL
NET: ------------------ NETBIOS Name Recognized ------------------
NET: Header length = 44, Data length = 0
NET: Delimiter = EFFF (NETBIOS)
NET: Command = 0E
NET: No LISTEN command outstanding for this name.

| 1 Help | 2 Set mark | 4 Zoom in | 5 Menus | 6 Disply options | 7 Prev frame | 8 Next frame | 10 New capture |

The interpreted detail of the frame highlighted in the summary window at the top of the screen.

Fig. 4-7. (Screen format © Network General Corporation.)

SUMMARY	Rel time	From IBM Portable		From Eagle One
44	0.000	NETBIOS	IBM Portable	NET Find name EAGLE <05>
45	0.477	NETBIOS	IBM Portable	NET Find name EAGLE <05>
46	0.977	NETBIOS	IBM Portable	NET Find name EAGLE <05>
49	1.477	NETBIOS	IBM Portable	NET Find name EAGLE <05>
50	1.977	NETBIOS	IBM Portable	NET Find name EAGLE <05>
51	2.476	NET Session alive		
53	2.577	NETBIOS	IBM Portable	NET Find name EAGLE <05>
59	7.582	NETBIOS	IBM Portable	NET Find name EAGLE <03>
60	7.593			NET Name EAGLE <03> recognized
61	7.603	NET D = 02 S = 04 Session initialize		
63	7.613			NET D = 04 S = 02 Session confirm
65	7.629	NET D = 02 S = 04 Data, 70 Bytes		
67	7.642			NET D = 04 S = 02 Send more now
69	7.652	NET D = 02 S = 04 Data, 70 bytes		
71	7.667			NET D = 04 S = 02 Data Ack
73	7.713			NET D = 04 S = 02 Data, 35 bytes
75	7.723	NET D = 02 S = 04 Data ACK		

1 Help	2 Set Mark	5 Menus	6 Disply options	7 Prev frame	8 Next frame	10 New capture

In frame 44, the IBM Portable asks, "Is there a station out there receiving forwarded messages for the EAGLE?" The EAGLE is not having its messages forwarded, so no one replies.

Fig. 4-8. (Screen format © Network General Corporation.)

SUMMARY	Rel Time	From IBM Portable	From Eagle One
44	0.000	NETBIOS IBM Portable	NET Find name EAGLE<05>
45	0.477	NETBIOS IBM Portable	NET Find name EAGLE<05>
46	0.977	NETBIOS IBM Portable	NET Find name EAGLE<05>
49	1.477	NETBIOS IBM Portable	NET Find name EAGLE<05>
50	1.977	NETBIOS IBM Portable	NET Find name EAGLE<05>
51	2.476	NET Session alive	
53	2.577	NETBIOS IBM Portable	NET Find name EAGLE<05>
59	7.582	NETBIOS IBM Portable	NET Find name EAGLE<03>
60	7.593		NET Name EAGLE<03> recognized
61	7.603	NET D = 02 S = 04 Session initialize	
63	7.613		NET D = 04 S = 02 Session confirm
65	7.629	NET D = 02 S = 04 Data, 70 bytes	
67	7.642		NET D = 04 S = 02 Send more now
69	7.652	NET D = 02 S = 04 Data, 70 bytes	
71	7.667		NET D = 04 S = 02 Data ACK
73	7.713		NET D = 04 S = 02 Data, 35 bytes
75	7.723	NET D = 02 S = 04 Data ACK	

1 Help	2 Set mark	5 Menus	6Disply options	7 Prev frame	8 Next frame	10 New capture

Fig. 4-9. (Screen format © Network General Corporation.)

In frame 45, we see that nearly half a second has gone by (477 msecs), and the Portable tries again, with no better luck than the first time. In six tries, taking over 2.5 secs, there is no reply.

SUMMARY	Rel time	From IBM Portable	From Eagle One
44	0.000	NETBIOS	NET Find name EAGLE <05>
45	0.477	NETBIOS	NET Find name EAGLE <05>
46	0.977	NETBIOS	NET Find name EAGLE <05>
49	1.477	NETBIOS	NET Find name EAGLE <05>
50	1.977	NETBIOS	NET Find name EAGLE <05>
51	2.476	NET Session alive	
53	2.577	NETBIOS	NET Find name EAGLE <05>
59	7.582	NETBIOS	NET Find name EAGLE <05>
60	7.593		NET Name EAGLE <03> recognized
61	7.603	NET D = 02 S = 04 Session initialize	
63	7.613		NET D = 04 S = 02 Session confirm
65	7.629	NET D = 02 S = 04 Data, 70 bytes	
67	7.642		NET D = 04 S = 02 Send more now
69	7.652	NET D = 02 S = 04 Data, 70 bytes	
71	7.667		NET D = 04 S = 02 Data ACK
73	7.713		NET D = 04 S = 02 Data, 35 bytes
75	7.723	NET D = 02 S = 04 Data ACK	

| 1 Help | 2 Set mark | | 5 Menus | 6 Display options | 7 Prev frame | 8 Next frame | 10 New capture |

Finally, in frame 59, over 7.5 seconds into this sequence, Portable asks the Eagle directly (<03> instead of <05>), "EAGLE, can we get together for a chat?" - and gets an answer in 11 milliseconcs!

Fig. 4-10. (Screen format © Network General Corporation.)

125

The received test pattern is then compared bit by bit with a locally generated pattern to detect the presence of errors. The repetition period of the test pattern must be selected to provide a sufficiently smooth spectrum for the system data rate. The recommended pattern lengths for the most common data rates are illustrated in TABLE 4-1.

Because the out-of-service measurement eliminates any traffic, it can only be used during production, installation, or for experimental testing.

In-service error measurements can sometimes be performed if the traffic has an inherent repetitive pattern, if the line format has some inherent error detection, or if the received signal can be monitored for certain threshold crossings. In-service techniques are an estimate of the error rate and might not provide a true indication. They can be used for performance monitoring during system demonstration.

A typical test configuration for BER measurements is shown in FIG. 4-11. The receiver can measure error pulses, average BER, errored seconds, or errored blocks. Forced errors can be injected at the transmitter and observed at the receiver for an indication of loss of synchronization.

Error-free seconds are another type of error measurement. They are expressed as the percentage or probability of seconds that are error free in a measurement period. The measurement can be done in either a synchronous or asynchronous mode.

In synchronous mode, an errored second is defined as a second interval after the occurrence of the initial error. The synchronous mode allows measurements to be made with different instruments on the same link. The measurement does not provide error-free seconds directly, but it is a measure of error-free time.

In the asynchronous mode, each discrete 1-second interval is tested for errors. This measurement provides error-free seconds directly.

Table 4-1. Pseudorandom Sequence
Length for the Measurement of Error Rate.

Bit Rate	Pattern Length	CCITT Recommendation
Up to 20 kb/s	$2^9 - 1$	V.52
20 to 72 kb/s	$2^{20} - 1$	V.57
1.544 Mb/s	$2^{15} - 1$	O.151
2.048 Mb/s	$2^{15} - 1$	O.151
6.312 Mb/s	$2^{15} - 1$	O.151
8.448 Mb/s	$2^{15} - 1$	O.151
32.064 Mb/s	$2^{15} - 1$	O.151
34.368 Mb/s	$2^{23} - 1$	O.151
44.736 Mb/s	$2^{15} - 1$	O.151
139.264 Mb/s	$2^{23} - 1$	O.151

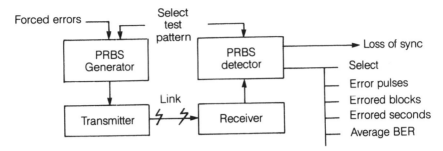

Fig. 4-11. BER measurement.

The use of asynchronous errored seconds is advised in CCITT Recommendation G.821 while other standards in use in North America advise the use of synchronous errored seconds.

Bit error-rate measurements require the use of a confidence level. This is the probability that the measured error rate is within an accuracy factor of the true average BER. Stated in terms of the number of errors:

$$\text{Confidence level} = P(np \leq ak_i)$$

where

$\quad\quad n$ = number of trial bits
$\quad\quad p$ = probability of bit error
$\quad np$ = expected number of errors
$\quad\quad a$ = accuracy factor
$\quad\quad k_i$ = number of measured errors

A confidence level model can be based on the Bernoulli trials model if the errors are independently distributed. Assuming this independence, the probability that ak_i or fewer errors will occur in n independent trials becomes

$$P\ (<ak_i \text{ errors in n bits}) \ ^{3/4} \sum_{k=0}^{ak_i} \binom{n}{k} p^k q^{n-k}$$

where

$$q = 1 - p.$$

This expression is difficult to evaluate for n, but approximations can be made for two important cases. As n becomes large, the expected number of errors also becomes large.

Then $np >> 1$, and the approximate expression becomes:

$$P\ (\leq ak_i \text{ errors in n bits}) \ = 1 - \text{erfc}\ [(a-1)\sqrt{k_i}] \quad\quad\quad (np >> 1)$$

The complementary error function is denoted by erfc (x).

If p is small and n is large, then the expected number of errors is small $(np \approx 1)$ and the probability approaches a Poisson distribution and can be expressed as:

$$P(np \leq ak_i) = 1 - e^{-ak_i} \sum_{k=0}^{k} \frac{(ak_i)^k}{k!} \qquad (np \approx 1)$$

These expressions can be solved to find the probability (confidence level) as a function of the errors measured (k_i) and the actual bit error rate (np).

If the probability is fixed, the BER can be computed as a function of the measured error for some specific confidence limit.

Plotted curves for 99 percent C.I. indicate that the actual BER is less than np/k times the measured BER. If ten errors are recorded, then the actual BER is within two times the measured BER with a 99 percent confidence.

JITTER

Jitter is a measure of the short-term variations of the parts of a digital signal from their ideal position in time. Jitter can be characterized in several ways: tolerable input jitter, output jitter in the absence of input jitter (also called intrinsic jitter), and the ratio of output to input jitter (also called the *jitter transfer function*.

The basic technique is shown in FIG. 4-12. The measurements are done out-of-service and are often combined with factory or system acceptance testing. The sources for input jitter testing are connected as shown.

The clock signal produced by the frequency synthesizer is modulated by the jitter generator and clocks the pattern generator. The output of the pattern generator is sent to the unit under test. The pattern detector monitors the output signal and provides an error output to a counter.

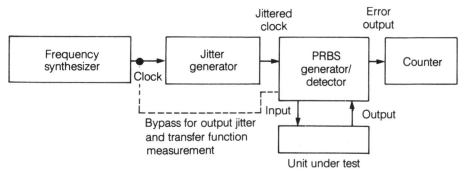

Fig. 4-12. Jitter measurement.

The amplitude of the induced jitter is adjusted until bit errors are detected, and this test is then repeated for a range of jitter frequencies. A plot of maximum tolerable input jitter versus jitter frequency is often useful.

Intrinsic jitter and the jitter transfer function are measured in a similar way except that a jitter detector replaces the error detector. Output jitter is measured in the absence of input jitter by circumventing the jitter generator with a jitter-free clock generating the PRBS pattern. The received pattern is applied to the jitter detector, which measures jitter amplitude.

The jitter transfer function can be found by using a PRBS signal modulated by the jitter generator and then measuring the gain or attenuation of jitter in the detected PRBS signal.

In both of these tests, the jitter amplitude, which can be rms or peak-to-peak, is measured for the bandwidths of interest. Bandpass, low-pass, or high-pass filters can be necessary to limit the jitter bandwidth. CCITT Recommendation 0.171 provides information for specific bandwidth measurements.

BIT-COUNT INTEGRITY

This refers to the preservation of the number of bits, characters, or frames that originated in a message. Bit-count integrity can be dependent on several parameters, such as signal-to-noise ratio, clock accuracy, and clock stability. Testing is accomplished by observing the conditions under which BCI is lost and then regained. Testing is done out-of-service. It is often a part of factory tests or system installations.

For example, a multiplexer usually is specified for a certain error rate. In order to test for the error rate, the multiplexer data stream is subjected to the specified errors while the time from the loss of BCI and to the regaining of BCI is measured. The measured times can then be compared to the specification.

RELIABILITY

Hardware reliability can be verified with a combination of burn-in testing, analysis, and field results. Burn-in tests can be used to verify the mean time between failure (MTBF) specification (see Appendix).

Using a number of units under test at one time, it is possible to demonstrate the reliability in a shorter time using sequential testing. The cumulative number of failures is plotted versus the cumulative test time until an accept or reject decision is reached, as shown in FIG. 4-13 or until the maximum allowable test time is reached.

For redundant equipment, direct testing usually is not practical for verifying the mean time between outage (MTBO) because of the time that would be required. The usual method of verifying the MTBO is to use analysis and MTBF testing. The analysis provides a list of failure modes and effects. It can

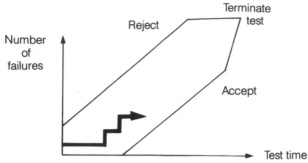

Fig. 4-13. Accept/test plot.

also be used to verify that performance sensors will detect failures, and that the switching circuits will cause the switchover at the proper time from failed units to standby units. The MTBO can then be calculated from a demonstrated MTBF and the predicted redundancy switching.

Reliability also can be monitored and recorded under actual field conditions. High-failure items can be identified and replaced before actual failure occurs.

LINK TESTING

Before the acceptance of a cable link, testing is often performed to ensure that the link installation meets specifications. An important part of this is the verification of equipment interface compatibility and operation in a typical environment.

The equipment can be tested in a loopback mode as discussed earlier rather than over the link. The test period should be long enough to identify cable mismatches, faulty grounds, and other defects.

Another part of link testing is the actual operation of the transmission equipment over the link. Parameters such as signal-to-noise rations usually can be measured quickly. Error rates can be measured for digital data, and voice-channel measurements can be made for 4-kHz analog signals.

The link also can be stressed to verify the design margins. This is done by first determining the nominal received signal level (RSL). Then a BER test is conducted with a calibrated attenuator inserted in the received signal path and set to yield the threshold BER. The margin should be determined separately for each received path.

Hysteresis can be tested by fixing the attenuation of one branch at a constant value and then varying the attenuation of a second branch above and below the design value. If the measured margin and hysteresis are almost equal, the link can be expected to meet performance objectives.

PERFORMANCE MONITORING

The threshold characteristics of digital transmission require a different approach in determining performance when compared to analog transmission. An analog signal will suffer a gradual effect that is transformed into degraded performance. The gradual change can easily be noticed. The gradual degradation of a digital signal has no noticeable effect on system performance until errors begin to appear. When a digital signal has deteriorated to the point where errors begin to be noticed, only a small margin is left between this initial point and unacceptable performance. An effective digital-performance monitor should be able to constantly measure the margin between the state of the transmission channel and the threshold of error of that channel. The performance should also allow the operator to implement diagnostic and maintenance procedures before the actual user performance degrades.

Performance monitors should also have the following general characteristics:

- The resolution should be great enough to allow a continuous quantitative measurement of the margin.
- In-service methods must be used to allow normal traffic without affecting the end user.
- The dynamic range should extend from threshold to received signal levels that are significantly higher than the threshold.
- There should be a rapid and observable response to any system degradation.
- A functional or empirical relationship with the BER should exist.
- Simplicity should be emphasized to maintain reliability and minimize cost.

Several basic hardware-performance-monitoring techniques are available. These are described in the following section.

TEST-SEQUENCE TRANSMISSION

When channels are free in a multiplexed configuration, these can be used as test channels. Auxiliary service channels of this nature are often used for the transmission of a pseudorandom binary sequence at a bit rate.

The test sequence is interleaved with normal traffic at some fraction (usually 1/n) of the system bit rate. At the receiver, the BER can be estimated using the ratio of errors detected to the number of test bits transmitted. The overhead for this technique is a function of the time required to recognize a specified error rate multiplexed by n times that required to count all errors for the transmitted bits. This is usually a relatively long time, even for low error rates.

FRAME BIT AND PARITY-BIT MONITORING

Digital multiplexers use a repetitive frame pattern for synchronization. The pattern is fixed and known beforehand, so an estimate of BER can be made by measuring the frame-bit errors. Performance monitoring can be accomplished with the multiplexer's digital pulse output, which occurs for each frame-bit error. A counter is used to determine the frame-bit error rate. Some provide a front-panel display of error rate. The required counting and averaging are done internally.

Parity or *check bits* can be added to blocks of data or to frames in order to provide error detection. At the transmitter end, a parity bit is added to each block of data to make the sum of all bits always odd or always even. At the receiver end, the sum of the bits is computed in each block or frame and compared to the value of the received parity bit. A difference between transmitted and received parity bits indicates that an error has occurred. Parity checking is a useful estimate of system error rate for error rates below 10^{-n}, where n is the length of the block or frame.

Several multilevel techniques place constraints on transitions from one level to another. A received signal can be monitored for differences from these level constraints to provide error detection. The level constraints depend on the code used. A bipolar code should not have two consecutive marks with the same polarity.

EYE PATTERN MONITORING

The eye pattern opening (FIG. 4-14) provides a qualitative measure of error performance. As the signal degrades, the eye pattern becomes blurred, and the opening between levels at the symbol sampling time becomes smaller. Noise and amplitude distortion reduce the size of the vertical opening while jitter and timing changes reduce the size of the horizontal opening.

This reduction or closure can be compared with values measured with the system operating in an ideal environment. The closure can be expressed in decibels by equating the system degradation change to $-20 \log (1 - \text{eye closure})$. This can be converted to a BER estimate by a calibration to the eye pattern change.

A more general technique is to analyze selected bit patterns. The received eye pattern is compared to the ideal eye pattern as discussed above for a selected bit pattern.

PSEUDOERROR DETECTION

This technique is based on degrading the receiver margin intentionally. The pseudoerror is implemented by offsetting the sampling instant and/or the decision threshold from their optimum values and counting the number of detected

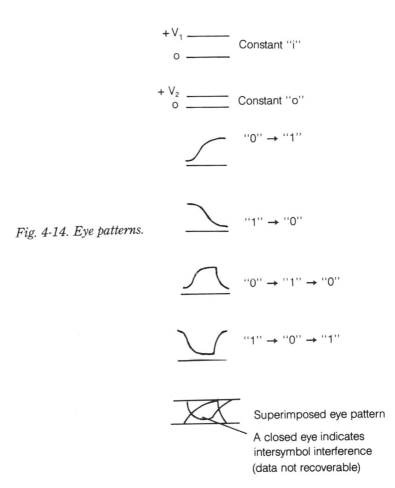

+V₁ ——— Constant "i"
o ———

+V₂ ——— Constant "o"
o ———

"0" → "1"

Fig. 4-14. Eye patterns.

"1" → "0"

"0" → "1" → "0"

"1" → "0" → "1"

Superimposed eye pattern

A closed eye indicates
intersymbol interference
(data not recoverable)

bits (the pseudoerrors) in the offset region. For each type of degradation, the relationship between the actual BER and pseudoerror rate (PER) is fixed. This relationship depends on the amount of intentional offset selected for the pseudo-decision circuit. The PER is typically greater than the BER and can be estimated in a shorter time, especially for low error rates.

The pseudoerror technique is preventive in nature because it detects channel degradation before actual errors occur. The calibration of PER to BER should be related to the potential sources of degradation for the application.

A comparison of BER and PER as a function of h indicates that over a range of zone heights the PER is almost parallel to the BER. The pseudoerror rate at h = 0.2 is about two orders of magnitude greater than the ideal error rate.

ERROR CONTROL

A basic way of improving reliability in a network is to add error-control processing at the transmitting and receiving ends. This is normally done through the channel encoder by adding redundancy in such a way that the decoder decision can be based on several received symbols rather than just one.

The particular error-control approach implemented depends on the performance expected of the approach and the channel environment in which it must operate. Other relevant implementation issues include the required information throughput rate, the allowed throughput delay for channel encoding and decoding, modem modulation, symbol-rate, tracking-accuracy, and phase-ambiguity.

Phase-ambiguity is important because the modem can lock up to a signal that is 180° or ±90° out of phase. Differential encoding and decoding can be used with some codes, but performance usually is better when the differential coding is placed outside the channel coding. The channel codes can also be selected so that the decoder resolves phase ambiguities without differential coding.

Error control can also use soft or hard receiver quantization. Consider a system that uses binary phase-shift keying (BPSK) modulation. If hard receiver quantization is used, the demodulator will instruct the decoder if the demodulator output for each symbol was positive or negative. If soft receiver quantization is used, the demodulator also provides additional bits of information based on the degree of confidence the demodulator has in the sign decision. This confidence level is a function of the magnitude of the demodulator output.

The acceptable bandwidth expansion is important because the required channel encoders add redundancy and for a fixed source symbol rate, this means a higher channel symbol rate and generally, a larger bandwidth requirement. In a bandwidth-limited system, only a small amount of redundancy is possible. A selective approach sometimes can be used with more redundancy in some links.

The following characteristics of the bursts on bursty channels are important:

- Duration and attenuation
- Periodic or random characteristics of the duty cycle
- Guard space between bursts

The burst rate and duty cycle of the received data determine the suitability of using a decoder at some smoothed average rate rather than at the burst rate. Other important characteristics include:

- Required decoder synchronization and loss-of-sync times
- Required bit, symbol, or message-error probability or error rate

- Allowed rate of accepting incorrect messages
- The availability and characteristics of return channels
- The vulnerability of the destination data to bursts of errors from the decoder
- The number of separately encoded sequences that must be simultaneously decoded
- Mutual-user or hostile jamming

CODING TECHNIQUES

The most common error control techniques are:

- Forward-error control (FEC)
- Automatic-repeat-request (ARQ)
- Hybrid FEC-ARQ

In forward-error control, the decoder corrects as many channel errors as possible. The transmitted sequence is tested based on hard or soft-decision inputs.

In the ARQ automatic-repeat-request control, the decoder detects the errors and requests that the incorrectly received frame be sent again. The request can be repeated until the frame is received correctly. The repeated transmissions tend to reduce the throughput rate of the system.

The throughput efficiency of an ARQ system is equal to the ratio of the total number of information bits accepted by the decoder per unit time to the total number of bits that could have been transmitted in the same amount of time without ARQ. This ratio depends on the channel error. ARQ protocol, data rate, propagation delay, and frame length.

The three basic types of ARQ protocols are:

- Stop-and-wait
- Go-back-n
- Selective-repeat

In stop-and-wait ARQ, the source sends one frame at a time and waits for an acknowledgment from the destination that the frame was accepted by the error-detection decoder.

When the acknowledgment is received, the next frame is sent. If the error-detection decoder at the destination detects an error, it provides a negative acknowledgment, and the source terminal repeats the frame. A half-duplex channel can be used because information is sent in only one direction at a time. Although the technique is simple, it tends to be inefficient due to the waiting time for acknowledgments that are required. A larger frame length can reduce the idle-time inefficiency, but a longer frame length increases the chance that the frame will have an error.

In continuous ARQ, the source sends frames continuously without waiting for an acknowledgment that the frame was accepted. When the source terminal goes back a number (n) of frames and transmits the frames again from the point where an error was reported, then we are using the go-back-n technique. It can be more effective than to the stop-and-wait technique, but it becomes inefficient when n becomes large. This problem can be avoided by retransmitting only the specific frames that were missed, using a selective-repeat approach.

ARQ systems are helpful in applications in which channel errors are likely to occur in bursts. If random errors are present, the throughput efficiency will drop quickly at higher channel error rates. A hybrid ARQ might be used in which FEC coding is used to correct the randomly distributed errors and an ARQ scheme is used to reduce the error rate.

BLOCK AND CONVOLUTIONAL CODES

FEC codes can be either block or convolutional. *Block codes* process data in blocks that are independent of one another. A block encoder accepts data in blocks of k symbols, and provides output encoded data in blocks of n symbols. Because redundancy is added, n will be greater than k. The ratio k/n is called the *code rate*, or rate (R).

The code rate differs from the data rate of the code. The data rate is the rate at which information is presented to the encoder. The code rate out of the encoder is also called the *channel data rate*. It is 1/R times larger than the data rate. Both usually are given in bits per second (bps).

Convolutional codes provide n output symbols for each group of k input symbols. The output depends not only on the last set of k input symbols but also on several of the preceding sets of input symbols.

In convolutional codes, k and n are smaller than in block codes. The code rate is still k/n.

The reciprocal of the code (n/k) rate is called the *bandwidth expansion* because adding coding increases the required channel symbol rate. Codes with low rates generally provide greater error detection and correction, but they require a larger bandwidth expansion.

Linear block codes can be viewed as a k × n generator matrix G. In the binary case, all the elements in the generator matrix and the encoder input and output vectors are "0" or "1." The arithmetic is performed using ordinary multiplication rules and bit-by-bit modulo-2 addition.:

$$0 + 0 = 0$$
$$0 + 1 = 1$$
$$1 + 0 = 1$$
$$1 + 1 = 0$$

$$0 \ . \ 0 = 0$$
$$0 \ . \ 1 = 0$$
$$1 \ . \ 0 = 0$$
$$1 - 1 = 1$$

Because

$$1 + 1 = 0$$
$$1 = -1$$

subtraction = addition

The code words are bit-by-bit modulo-2 sums of the rows of G. In order to guarantee that there are two distinct code words, the rows of the matrix G must be linearly independent.

Linear block codes are often decoded using algebraic or table-lookup techniques based on hard-decision demodulator outputs. Soft decision implementations are generally used with convolutional coding techniques. Block codes are often used on channels in which only hard receiver decisions are available and the data is already in a blocked format. Another disadvantage of block codes compared to convolutional codes is the larger n-way ambiguity, which must be resolved to determine the start of a block.

CYCLIC CODES

Cyclic codes are a class of linear block code in which each end-around cyclic shift of each code word is also a code word. The generator and parity-check matrices of cyclic codes can be expressed in terms of the coefficients of the generator and parity polynomials. For encoding a cyclic code, a multiple-stage feedback shift register can be used with the feedback tap multipliers based on the coefficients of the generator polynomial.

Cyclic codes with their cyclic properties require relatively simple encoder circuits. Cyclic codes are typically employed in error detection.

The cyclic-redundancy-check (CRC) code is a common error-detecting method. This type of code uses CRC error burst (of length B) in an n-bit received code word. The error burst might appear as a contiguous sequence, or as an end-around shift of a contiguous sequence. The first and last bits and any number of intermediate bits are received in error. TABLE 4-2 shows the error-detecting capabilities of CRC codes.

Cyclic codes often have a very large block length (n). This code can be shortened by deleting the first few information bits. The result is no longer a pure cyclic code but a modified code that is still referred to as a CRC code. All standard CRC codes are of this type and have the same minimum distance and error-detection properties as the pure cyclic code.

Table 4-2. CRC Error—Detecting Capabilities.

For a block code with n channel bits d(min) minimum distance and k information bits

Detectable N-Bit Channel-Error Patterns

All CRC error bursts of length n − k or less
All combinations of $d_{(min)}$ − 1 or less errors
All error patterns with an odd number of errors when the generator polynomial has an even number of nonzero coefficients.
$1 - 2^{-(n-k-1)}$ of the CRC error bursts for length B = n − k + 1
$1 - 2^{-(n-k)}$ of the CRC error bursts for length B > n − k + 1

The encoder for these shortened CRC codes also can be implemented by computing the end-around shift and passing the block of data after checking. One basic approach is to pass the block of data when the computed parity bits are all equal to the received parity bits.

The polynomials for the most common CRC codes are shown in TABLE 4-3.

HAMMING ERROR-CORRECTING

The *Hamming code* is an example of a binary linear block error-correcting code. Hamming codes are able to correct all single errors or all combinations of two or fewer errors. They belong to a type of linear block codes known as *perfect codes*. A perfect code for correcting x bits has the characteristic that every error pattern of x or fewer errors can be corrected, and no error pattern with more than x errors can be corrected. Hamming codes have the parameters of TABLE 4-4.

EXTENDED HAMMING CODE

If a parity bit is added to the Hamming code, then an *extended Hamming code* is formed with the parameters of TABLE 4-5.

Code	Polynomial
CRC-12	$p(x) = 1 + x + x^2 + x^3 + x^{11} + x^{12}$
CRC-16	$p(x) = 1 + x^2 + x^{15} + x^{16}$
CCITT CRC	$p(x) = 1 + x^5 + x^{12} + x^{16}$

Table 4-3. CRC Codes.

n	k	m	d(min)
$2^m - 1$	$2^m - 1 - m$	n − k	3
			3,4,5,6 . . .

where:
    k = input block length
    n = output block length

Table 4-4. Hamming Codes.

Table 4-5. Extended Hamming Code.

n	k	m	d(min)
2^m	$2^m - 1 - m$	$n - k - 1$	4

This code is able to correct any single error and double errors or any combination of three errors.

Normally, table lookup is used for decoding these codes.

GOLAY CODE

The Golay code is another perfect code. It has a minimum distance of seven and is able to correct any combination of three or fewer errors or detect any combination of six or fewer errors. Two polynomials that can be used to generate the Golay code are shown below:

$$p(x) = 1 + x^2 + x^4 + x^5 + x^6 + x^{10} + x^{11}$$
$$p(x) = 1 + x + x^5 + x^6 + x^7 + x^9 + x^{11}$$

An extended Golay code also can be formed by adding a parity bit to the basic Golay code.

Generally, the minimum distance of any binary linear block code with an odd minimum distance can be increased by one by adding an overall parity bit.

CONVOLUTIONAL CODES

Convolutional codes are based on the property that each output group depends not only on the k inputs at that time but also on a number of preceding input groups. These codes need to store the input groups. They accept data in small groups of k symbols and provide the encoded output data in small groups of n (n > k) symbols.

Convolutional codes allow decoding algorithms that can be used with soft receiver decisions, and they are popular because they are simple to implement. The encoder provides each of the k input sequences directly as outputs as well as n − k parity output sequences.

There is no need to send known tail bits when blocks of data are transmitted, so the tail sequence or trailer can be shorter.

Convolutional codes normally use encoders based on feedforward logic. Encoders with feedback logic also can be used, but these have a higher bit-error rate than an equivalent code generated using only feedforward logic.

VITERBI DECODING

Viterbi decoding takes a maximum-likelihood decoding approach, using a convolutional code that selects the most likely encoder input sequence based on

Table 4-6. Viterbi Decoder Performance.

Bit Error Rate	Bit Energy-to-Noise Density (dB)		
10^{-3}	3-9	6.9	
10^{-4}	4.8	8.4	Modem
10^{-5}	5-5	9.6	without
10^{-6}	6.0	10.6	coding
3-bit soft quantization, R = ¾			

the particular received sequence. The path structure takes the form of a trellis or tree representation, and the Viterbi decoding algorithm moves through testings as a progression of each received sequence with a hard-decision decoder. Typical performance specifications for this code are shown in TABLE 4-6.

The path branches are tested for the number of bits in which the received symbols differ from the encoder output symbols. The decoder compares these conditions for the two paths entering each node and eliminates the path with the largest number of differing bits.

SEQUENTIAL DECODING

Viterbi decoders check all possible state transistions at each step, with the implementation complexity increasing almost exponentially with increasing length. Sequential decoding is a decoding approach that checks only some of the branches.*

The number of branches checked depends on the channel characteristics. When a period of reliable data is received, only a few paths are checked. When a sequence of unreliable data is detected, many more paths are checked.

The sequential decoder systematically searches through the coding tree and extends one branch at a time. It bases its decision on which path to add on tests of the already examined paths.

Typical performance specifications of a 2-bit soft-quantization decoder are shown in TABLE 4-7.

FEEDBACK DECODING

Feedback-decoded convolutional coding systems use hard-quantized received data to compute a sequence in a way similar to that for block codes. Some of the bits for every received branch are used to address a ROM whose output bits then provide an estimate of the errors in the data portion of an earlier branch of received bits.

---

*See *Principles of Digital Communication and Coding*, by A. J. Viterbi and J. K. Omura, New York: McGraw-Hill, 1979.

Table 4-7. Sequential Decoder Performance.

Bit Error Rate	Bit Energy-to-Noise Density (dB)			
	100 kbps	500 kbps	1.54 Mbps	12 Mbps
10^{-5}	4.9	5.2	5.4	5.7
10^{-6}	5.5	5.7	5.9	6.1
10^{-7}	6.0	6.2	6.4	6.6

R = ⅞, BPSK or QPSK Modulation

When the ROM output indicates errors have occurred, the errors are removed from the delayed data copy and the sequence is updated as a result of the errors.

The use of feedback can result in unlimited error propagation because a finite number of errors can cause the decoder to make an infinite number of updates. This occurs when the number of ROM address bits is not large enough.**

Feedback decoding is simple to implement; however, it does not provide all of the coding benefits that can be obtained with Viterbi or sequential decoding. Interleaving and deinterleaving can easily be included as part of the encoding and decoding processes. Internal interleaving and deinterleaving can be achieved by using a shift register for each delay element.

Interleaving allows the errors in a burst to be treated as isolated bit errors. A channel error burst is divided so that the bits used in the table lookup are based on only one channel bit error.

The feedback decoding technique also can be used with cyclic block codes. These types of decoders are sometimes called *Meggit decoders*.

In some systems, the table look-up decision device can be replaced with a threshold device. This technique is known as *threshold decoding*.

BURSTY CHANNELS

Error-correcting decoders use a group of received channel symbols to estimate the original information sequence. Good decoder-error-rate performance occurs if there is a high percentage of reliable symbols in the received group that the decoder bases its estimate on. In bursty channels, a large portion of the sequence can be unreliable. FEC or ARQ techniques can be used in this type of environment.

ARQ uses an error-detecting code to detect blocks or frames of data with errors and requests that the incorrect frames be repeated.

**See *Error-Correction Coding for Digital Communications*, by G.C. Clark, Jr. and J.B. Cain, New York: Plenum Press, 1981.

Error detection along with forward error control (FEC) can be used to correct as many errors as possible, which reduces the retransmission required. Some FEC codes are designed specifically to correct bursts of errors. These codes usually operate on hard-decision channels.

If an error burst of length is L in which the first, last, and intermediate symbols are received, then for (n,k) linear block codes, the number of parity-check symbols in a code capable of correcting all bursts of length L or less is bounded by

$$n - k \geq 2L$$

The efficiency of a block burst-error-correcting code is then

$$\frac{2L}{n - k} \leq 1$$

In a convolutional code, the burst-error-correcting efficiency also depends on the number of correctly received channel symbols required between the error bursts. *Rate (R) convolutional codes* are able to correct all bursts of length L or less, when there is an error-free guard space of length G on both sides of the error burst. Then

$$\frac{G}{B} \geq \frac{1 + R}{1 - R}$$

and the burst-error efficiency of the convolutional code is

$$\frac{B}{G}\left(\frac{1 + R}{1 - R}\right) \leq 1$$

Codes are available that allow burst-error-correcting efficiencies of close to one.

Interleaving FEC Codes is often a good approach in a bursty environment, especially when soft-received decisions are used. The order of the transmitted symbols is changed such that long bursts are not sent to the decoder. In this case, a random-error-correcting code is employed. The interleaving portion requires symbol reordering or permutation of the encoded symbols before transmission. An inverse operation is required at the receive terminal, which reorders the permuted sequence of symbols so that it corresponds to the original encoder output ordering. This is referred to as *deinterleaving*. Generally the deinterleaving is done on soft-decision data, and the interleaving and deinterleaving are done external to the encoding and decoding operations. This is often the case with feedback-decoded convolutional codes using hard-decision decoders, which employ interleaving external to the encoding and decoding operations.

Block interleaving views the interleaver memory space as a rectangular array of storage locations. Data usually is written in by column and when the array is full, and read out by rows. Deinterleaving is done with an inverse operation by writing the received input symbols into the rows of another array and reading them out by columns when the array is full.

Both arrays are used when data is arriving continuously. Symbols are written into one array while they are being read from the other.

BLOCK AND CONVOLUTIONAL INTERLEAVING

Interleaving must provide a separation between any two symbols from the burst that is greater than or equal to the memory span of the decoder. For linear block codes, the decoder memory span is equal to the block length of the code. If the memory span is n, then there must be at least (n-1) reliable symbols, not from the burst, but between any two unreliable symbols from the burst.

The most common variations of convolutional interleaving are synchronous, Ramsey, and helical interleaving. Functionally, a series of commutators are placed on the input and output of the interleaver. These are synchronized with each other, and after each symbol is shifted in, they are sent to the next branch. The deinterleaver commutators are synchronized with each other in a similar way.

The deinterleaver disperses the unreliable channel symbols such that any two channel symbols in a contiguous input burst are separated by symbols at the output.

When the deinterleaver commutators are synchronized to those of the interleaver, there is a constant throughput delay of the channel-symbol time intervals.

Convolutional interleavers and deinterleavers are usually implemented with shift-registers or random-access memories (RAMs). The read and write RAM addresses are incremented by either 1 or some fixed integer amount. Each time a new symbol is received by the interleaver or deinterleaver, it is written into the RAM, and a symbol is read out.

For convolutional decoders, the memory-span requirement depends on the code rate. Applications in which only small error-rate increases caused by bursty channel conditions can be tolerated require high memory span capability. Other considerations are throughput delay and the synchronization time.

Throughput delay is the delay from the time a symbol is written into the interleaver until it is read out of the deinterleaver given a zero channel delay.

Deinterleaver synchronization requires that the received symbol sequence be in accordance with the deinterleaver array mapping. In block interleaving, the deinterleaver must know which of the memory array locations should be used to store the first symbol. This usually is done with a short preamble. In order to reduce transmission overhead and implementation, the preamble can be written over the encoded data to be transmitted with the decoder correcting the inserted errors. This reduces the error-rate performance of the system, but it can be used in low-rate applications.

When continuous data is being transmitted, the decoder can be used to determine when the deinterleaver is out of synchronization and when an adjustment is needed. The decoder checks a number of quality (q) bits for synchronization. The number of deinterleaver synchronization states is a multiple of the block length for block codes or the number of channel symbols per branch (n) for convolutional codes.

Block interleaving is useful when the data is already in a block format or when a random approach is needed. When the data are received continuously or in long blocks, the double buffering required with block interleaving, as well as the large throughput delay and the number of synchronization states, tends to make a convolutional interleaving more suitable.

Block interleavers and deinterleavers are normally implemented with random-access memories (RAMs). The memory addresses in the array are assigned so that in interleaving the symbols are written into the RAM in a permuted manner and read out from a sequential sequence of addresses. In the deinterleaver, the symbols would be written into the RAM at a sequential sequence of addresses and read out in a permuted manner. These operations also can be reversed.

If the permuted sequence of addresses is randomized, the process is called *random-block interleaving*. This technique is useful in applications with intentional interference or unpredictable burst characteristics.

The size of the memory required will depend on the type of decision process used. If q-bit soft-receiver decisions (1 sign and $q - 1$ quality bits) are used, the deinterleaver memory requirement is multiplied by q.

5

Network
Security Techniques

Communication systems composed of networks of computers and users incorporate cooperative schemes such as resource sharing for the user community. The security of the information on a particular host then is dependent on the security measures employed. Many of the security techniques used for an onside computer system security are applicable or adaptable for use in the larger network environment.

INTRODUCTION

Networks can range from collections of heterogeneous, autonomous hosts to groups of hosts operating under a single authority and cooperating to provide a coherent interface, as in the case of many distributed systems. The types of methods used for network security vary depending on the network environment.

The security mechanisms incorporated in a computer system accessed via a network can be ineffective if the network fails to provide a secure communication path between each user and the computer files. The interface between network and computer system security measures must be coherent across this interface and permit the interaction of security measures in a synergistic fashion. The design of the communication protocols, especially the end-to-end protocols, can be critical and requires careful coordination with the network security measures for the system to be both effective and efficient.

The adaption of the appropriate network security methods begins with an evaluation of the threat environment and an assessment of risks in that environment. The appropriate techniques are then selected for use in that particular environment.

As a general rule, security components that define and control each network element are easier to implement in a single-vendor network. In multivendor networks, a nucleus of large mainframes (such as IBM 30XX devices) might still exist, but surrounding it and connected to it are any number of boxes by various other vendors.

At the other ends of the communication lines (whether they are microwave, satellite, fiberoptic or cable) might be more non-IBM equipment. An ever-increasing variety of terminals and personal computers can be connected to these devices in many different ways.

Potential security violations involve the unauthorized release or modification of information and the unauthorized use or the denial of use of resources. Breaches of security related to the release of information correspond to passive wiretapping, while security breaches of the other form correspond to active wiretapping.

Unauthorized entry implies that the potential violation takes place contrary to security policy. The intruder could be a wiretapper outside of the user community or a legitimate user of the network. Communication security techniques are normally used to counter outside attacks. Authentication and access control are employed for the more difficult protection required in the case of legitimate users.

The following design principles are often employed in the implementation of secure systems:

- Economy of the mechanism
- Separation of privilege
- Lease privilege
- Least-common mechanism
- Fail-safe errors or defaults
- Complete mediation
- Open design concept

Economy of mechanism relates to the use of the simplest possible design to achieve the desired effect. Protection mechanisms should be kept simple because implementation errors that allow unintended access might not be detected during normal operational use.

Fail-safe errors require that the access decisions be based on permission instead of exclusion. Total mediation requires that each access be tested against an access-control database. Implementing fail-safe measures requires the selection of those objects that should be accessible while the mediation requirement enforces a system-wide access control. These requirements tend to complement each other.

An open design implies that the design of the protection mechanisms should not be kept secret. The open design philosophy is based on security mechanisms that do not depend on the ignorance of potential attackers, but on the possession of specific, more easily protected keys or passwords.

The tenant of open design is also open to controversy. A design that is hidden tends to raise the cost of penetration, but this approach generally is not acceptable in a common-user-network environment.

The separation of protection mechanisms from protection keys allows the mechanisms to be reviewed by potential users without prompting concern that the review itself will compromise the safeguards. This becomes especially important in a network environment, where there is often a large and diverse group of users.

The tenant of separation of privilege states that a protection mechanism requires two keys for access. This technique is used commonly for bank safe-deposit boxes. Separation can be applied to the distribution and use of encryption keys, authentication procedures, and the access control.

The use of least privilege means that each program and user of the system operates with the least amount of privileges necessary to perform the required task. This tends to reduce the amount of damage that can result from an accident, error, or default in the system. An example of this is the military security rule of the need-to-know. Least privilege is important in the implementation of connection-oriented security measures. It commonly involves the use of selective encryption techniques.

The least-common mechanism of security implies that the amount of mechanism common to more than one user and depended on by the rest of the users be minimal. Each shared mechanism is a potential information leakage path and should not compromise security. The least-common concept affects connection-oriented security measures. The structure of the communications hierarchy involving encryption techniques is impacted by this concept.

SECURITY PHILOSOPHIES

When there are many different vendor products in a network, the data responsible for security must contend with various philosophies of security. Even if vendors classify security methods in great detail, they approach the subject differently, focusing on individual access and accountability, more control at the resource level, or more control down at the field level.

Other approaches include allowing control only down to the primary file or database without differentiating individual records or file members. Such systems use hierarchical levels of control, showing users a menu that indicates only what their individual passwords and clearance allow them to access in the system. Security can be added via an additional hardware or software product. These wares provide a basis for imposing the desired controls.

Security should be treated as a necessary overhead and designed into the system accordingly. For example, a system might use an override feature designed to limit access to sensitive information. Users should not be able to call up the override screen to give themselves access to sensitive information.

Data security can influence hardware and software acquisition decisions and can prevent weaker links from being chained into the network. Price, performance, and organizational constraints must be weighed, and data security will have to provide adequate safeguards despite the weaknesses of individual network components. Network developers and integrators, as well as auditors and data-security officers, must be cognizant of all the different vendor components in the organization's network. They must understand the security capabilities and the vulnerabilities of each components.

SECURITY APPROACHES

Communication network security techniques can be grouped into two basic approaches: link-oriented techniques and end-to-end techniques. *Link-oriented techniques* protect the message traffic independently on each communication link. *End-to-end techniques* provide uniform protection for protection for each message from its source to its destination. The two methods also differ in the nature of the security provided and the type of external interface presented to the users of the network.

With link measures, a break does not compromise the entire system. Idle nodes also can be fed a stream of unimportant ciphertext for camouflage purposes.

The use of end-to-end measures simplifies key distribution and creates fewer encrypting/decrypting operations, thus reducing the danger of a security break or errors. It also makes cost apportionment among users easier. End-to-end is often the best approach for most networks.

Link-oriented protection provides security for the message traffic passing over an individual communication link between two nodes. It is independent of the source and destination of the messages.

Link-oriented measures function with the data-link control protocols used such as SDLC and X.25 (level 2). The actual link might be the connection between a host and the communication subnet or between a terminal and the host to which it is connected.

It is generally easier to secure the terminal and host nodes than to secure nodes out in the communication subnet. When all subnet nodes are located near the hosts attached to them, they can be given the same level of security as the hosts. Employing link-oriented measures in the communication subnet has a disadvantage in that subversion of a subnet node can expose large amounts of message traffic. If the unsecured communication links lie within the communication subnet, link-oriented techniques are useful because they can provide a transparent communication security for the hosts attached to the network. If only some of the communication links between the terminals and hosts or between the hosts and communication subnet are required protection, security for these links can be implemented without affecting the other hosts on the network.

The cost of link-oriented measures within the communication subnet can be amortized over all of the users while the cost of link-oriented protection on terminal-to-host and host-to-communication subnet links can be assessed to the directly affected parties. In a common-user network, not all users might agree to the need for the security measures and might feel unjustly charged for this service.

END-TO-END SECURITY MEASURES

Rather than considering a network as a collection of nodes joined by links, which are independently protected, a network can be viewed as a medium for transporting messages from source to destination. End-to-end security measures protect the messages in transit between the source and destination nodes so that the subversion of any communication link between the source and destination does not result in exposure of the message traffic. End-to-end security measures can be implemented from host to host, terminal to terminal, and terminal to process.

End-to-end security often goes beyond the communication subnet, and it can require a higher level of standardization for the interfaces and protocols. An advantage of end-to-end security measures is that individual users and hosts can employ them without affecting other users and hosts.

CONNECTION-ORIENTED MEASURES

A communication network can be considered as a vehicle for providing users with connections or virtual circuits from source to destination. Security measures can be connection-oriented in which each connection or virtual circuit is protected on an individual basis.

Connection-oriented techniques can be thought of as a refinement of end-to-end measures. The same flexibility exists in selecting the points that act as the ends of the connection.

Connection-oriented techniques reduce the chance of undetected crosstalk. In many systems, they provide a higher degree of communication security. In a wide array of environments, they are more naturally suited to the user's concept of security requirements. This is because they rely on the security of equipment at the source and destination of a connection. In contrast, link-oriented techniques can require that each node in the communication subnet be secure. Both measures can sometimes be employed to provide a cost-effective hybrid system. This can offer a level of protection that is higher than can be achieved using either type alone.

WIRETAPPING METHODS

In order to enter a secured network, an intruder must gain access at some point in the network through which information will pass. In an internetwork system, this could be a gateway in some intermediate network that provides the communication path between processes. Even though the source and destination networks might be secure, access to the connection can be made as it passes through the gateway.

In passive wiretapping, the intruder observes the messages passing on a connection without interfering with their flow. This observation at the application level of messages is called the *release of message contents* and is the most common type of passive wiretapping. The observation can include parts of the message headers, and even if the application-level data is unintelligible, these header sections can indicate the locations and identities of the communication system.

An analysis of the lengths of messages and their frequency of transmission can be conducted with passive wiretapping. This type of traffic analysis can be used to gain knowledge of the nature of the data being exchanged. Traffic analysis can be blocked by using programs that modulate message destination, length, or frequency of transmission.

Active wiretapping involves some form of processing on the messages passing through the connection. The messages can be modified, duplicated, reordered, or inserted into the connection at a later time, or they might pass through unaffected. False messages can also be inserted into the connection. These acts are known as *message-stream modification*. If a message is discarded or delayed, this action is known as *denial of message service attack*.

The difference between message-stream modification and denial of message service attacks is a function of the degree of the attack and of the state of the connection. Different countermeasures are employed for each type of attack.

Connections should be initiated in such a way that secure identification of the users, terminals, and hosts at each end is possible and the integrity of the connection established. Verification of identity can involve both authentication and access controls beyond the level of the connection model, with a portion of the identification task occurring at the connection level.

ENCRYPTION

Encryption can be used as both a countermeasure to passive wiretapping and as a base for structuring countermeasures to active wiretapping. A *cipher* refers to the algorithmic transformation that is performed on a symbol-by-symbol basis on the data. There are technical differences between the terms encipherment and encryption, but the two terms are commonly used to refer to the application of a cipher to data.

An *encryption algorithm* is an algorithm that implements a cipher. The input to an encryption algorithm is referred to as *cleartext* or *plaintext*, and the output from the algorithm is called *ciphertext*. The transformation performed on the cleartext to encipher it is controlled by a *key*.

One line of encrypted text is the *authentication field*. If the message is a funds transfer, the authentication field might contain the number of dollars involved added to the date. A transfer of $2,842,000 transmitted on November 20 could have an authentication code of 3962: 1120 (the date) + 2842 (the amount in thousands of dollars).

Encryption systems employ software, or more commonly, hardware black boxes. The average box costs between $2000 to $4000, but systems costing as little as $400 per node are available.

Systems can be either on-line or off-line. On-line systems, the most common, sit between the terminal, or host, and the modem, or communications device. These provide end-to-end or link encryption.

Off-line systems use a device that converts cleartext into a ciphered message on a medium such as punched paper tape. The medium can then be delivered manually. Off-line encryption includes hand-held methods and microcomputer expansion boards.

To be useful, the encryption algorithm must be invertible, so there must exist a matching decryption algorithm that reverses the encryption transformation when used with the proper key.

In a conventional cipher, the key used to decipher a message is the same as the key required to encipher it. This type of key must be kept secret and is known only by users authorized to transmit and decipher messages using it. The concept of a conventional cipher is shown in FIG. 5-1.

In a public-key cipher, enciphering and deciphering are separate abilities. This is done using pairs of transformations, each of which is the inverse of the other and neither of which is readily derivable from the other.

This results in a multiaccess cipher in which one key is made public for use in enciphering messages for a given user, while the other is kept secret and used for deciphering messages sent to the user under the public key.

Key distribution and management requires some type of authentication. One basic technique that is used is the public key in which the sender encrypts the message first under the secret key and then under the public key of the intended receiver.

The receiver then strips off the outer layer of encryption using the secret key and completes the deciphering using the public key of the sender. The technique is shown in FIG. 5-2.

Fig. 5-1. Conventional cipher.

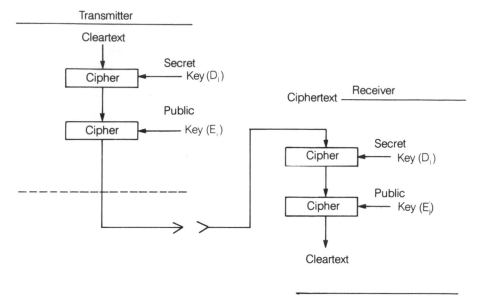

Fig. 5-2. Public-key cipher.

Public-key systems must use encryption and decryption transformations that are easy to compute and are inverses of each other. The key pairs also must be easy to generate and manage. It should not be possible to derive one transformation from the other or to invert the transformations without the correct keys.

Automatic key management is more expensive, but it allows the creation of keys as often as necessary. Keys are downloaded from the host computer on a regular basis.

When the key management is automatic, no one knows about it. It is essentially transparent, and no one is tempted to be dishonest.

Keys must be created, assigned, distributed and canceled in a way that minimizes the possibility of their falling into unauthorized hands.

The most common form of key distribution is sending the key physically to each site, normally in the form of a microchip sent by bonded messenger, certified mail, or other secure means.

Keys have to be changed regularly, and the receiving group should not know when the key will be changed. A more secure means of manual key distribution is to deliver only preliminary keys to different individuals. Two or more of the preliminary keys must be placed in the encryption device to generate the true key. In this way, the true key is never seen.

The major disadvantage of manual key distribution is that it becomes cumbersome when many terminals are serviced or when keys have to be issued daily or even more often.

Cryptosystems can be practically or unconditionally secure. A practically secure system is one where the computational cost required for penetration makes the task infeasible. An unconditionally or theoretically secure system is one that can resist any attack, regardless of the computational power and time required.

Cryptanalytic attacks can be classified into three hierarchical groups. Ciphertext-only attacks take place when the cryptanalyst uses a knowledge of the statistical properties of the language. The approach taken is to develop the probable words based on the relative frequency of letters or the use of parity bits in the transmission. Other parameters that can be used include the frequency of standard salutations in letters and common variable names in program text.

Known-plaintext attacks occur when the cryptanalyst matches plaintext with the ciphertext. This can occur in a network environment, because fixed-format messages are common with specific log-in greetings. There are often system-wide messages that can be matched.

Chosen-plaintext attacks extend this concept. The cryptanalyst is not only able to match plaintext and ciphertext, but also selects the plaintext. This can be attempted through the network mail.

A cipher in the network environment must be resistant to known and chosen plaintext as well as ciphertext-only penetration.

STREAM AND BLOCK CIPHERS

The two major types of encryption techniques in nonvoice telecommunications are *stream* and *block ciphers*. The block method enciphers fixed-sized blocks of bits using a key that is approximately the same size as the blocks being encrypted. The stream method uses bit-by-bit or byte-by-byte transformations on cleartext using a stream of key bits. This usually is done with an easily reversible operation such as addition by modulo.

Stream ciphers are structured to operate on the stream of cleartext in real time, enciphering each bit or byte as it is generated by combining it with a bit or byte from the key stream.

Most stream ciphers use pseudorandom bit streams with very long periods because of the typical volume of communications traffic and the logistic difficulties required to provide each user with enough key stream. These stream ciphers are susceptible to cryptanalysis.

A *Vernam cipher* is a stream cipher in which the key stream consists of a random bit string as long as the combined length of all messages that are ever transmitted using this stream. This type of cipher is theoretically and also practically unbreakable.

The source of the bits for the key system can be a pseudorandom-number generator that is started with a small initial key. It is completely independent of the cleartext stream, so changes to individual bits in the ciphertext do not propagate to other portions of the ciphertext stream.

Transmission errors can affect the value of bits of the ciphertext, but they do not change the ability of the receiver to correctly decipher subsequent transmissions. The undetected insertion or removal of bits from the ciphertext stream still results in a loss of deciphering capability. These ciphers are not often used because the lack of error propagation reduces the types of mechanisms available to detect length-preserving message-modification penetrations.

Autokey ciphers are a form of stream cipher in which the key stream is based based on the cleartext, ciphertext, or the key stream itself and an initial priming key. These ciphers can be used to detect errors in transmitted ciphertext in which errors will affect the correct decipherment of subsequent transmissions.

Ciphertext autokey (CTAK) ciphers use the transmitted ciphertext as input for key-stream generation. CTAK ciphers are self-synchronizing and recover from transmission errors after a fixed number of correct ciphertext bits are received.

Block ciphers transform blocks of bits under the control of a key. They map a set of cleartext blocks into a set of ciphered text blocks.

If the block size is n bits, the size of the cleartext space, which is the range of cleartext block values, and the size of the ciphertext space, which is the range of ciphertext block values is 2^n. The mapping must be invertible. The block cipher under the control of a single key requires the defining of one of 2^n permutations on the set of n-bit blocks.

It is normally not feasible to implement a block cipher that realizes all possible permutations because of the size of the key required and the complexity of the cipher that is also required. The block cipher usually is treated as a classical substitution cipher. If n is 7 or 8, the block corresponds to an individual character from a small alphabet.

This system is not very strong due to the small size of the blocks used. The cipher can be broken by comparing the frequency distribution of the individual blocks with the known frequency distribution of characters in large samples of cleartext.

The size of the block can be increased and the cipher constructed so that the frequency characteristics of the components of the block become hidden by mixing the transformations. The required frequency distribution and analysis then become very difficult.

DES

The Data Encryption Standard (DES) is the federal information processing standard. It is widely used for nonclassified government information and has become a de facto industry standard as well.

In 1986 the National Security Agency (NSA) announced that it would stop certifying network security equipment for compliance with the Data Encryption

Standard (DES). With the DES, one algorithm is used, and each user organization generates and distributes its own keys.

The NSA feared that because so much data is protected by one algorithm, a foreign power could concentrate resources on breadking DES. If the algorithm was broken, sensitive government and private sector data could be compromised.

The NSA decided that the way to reduce this vulnerability is to produce and distribute a number of algorithms from which users can choose and that it should control the keys. This would reduce the amount of data protected by any one algorithm.

The DES is a block cipher that uses 64-bit blocks and a 56-bit key. Each key defines a permutation on the space of 64-bit blocks. Each bit of ciphertext in a block is a function of each bit of the key and each bit of the cleartext block from which it was generated.

A change of one bit in either the key or the cleartext results in ciphertext in which each bit is changed with an approximately equal probability. A change in one bit either of the key or ciphertext will result in changes for an average of 50 percent of the bits of deciphered cleartext. This error propagation is limited to the block in which the error occurs, and decryption of other blocks is not affected.

This mode of DES is known as the *electronic code book* (ECB) because it acts like a conventional code book. The proper deciphering of messages is achieved when both sender and receiver employ the same key and the blocks are properly delimited. FIGURE 5-3 shows the operation of this mode of DES.

Because the DES is a conventional cipher, it can be used as part of a key-stream generator for a CTAK cipher. This is known as the *cipher feedback mode* of operation.

The DES is used as a self-synchronizing stream cipher that operates on cleartext strings of 1 to 64 bits in length. The cleartext is combined by modulo 2 addition with a matching number of key-stream output bits that are generated by the DES block cipher. Extra DES output bits are dropped.

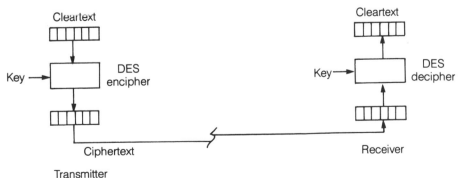

Fig. 5-3. Electronic code book DES mode.

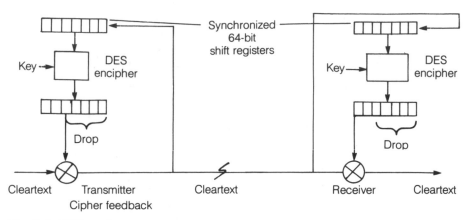

Fig. 5-4. DES mode.

The transmitted ciphertext is loaded into 64-bit shift registers that act as the input to the basic DES at both ends of the system. FIGURE 5-4 shows the operation of this mode operates with 8-bit bytes. The sender and receiver must use the same key, and both shift registers hold the same bit pattern.

When an error occurs in the ciphertext stream, a part of the received cleartext is lost but the receipt of 64 bits of error-free ciphertext will resynchronize the cipher. The number of bits actually affected by a single bit error is constant at 65 while the number of cleartext bits that might be useless by an error depends on both the length of the cleartext bit strings used and the degree of error detection used at higher-level protocols.

The DES is resistant to conventional cryptanalysis. Both ECB and CFB modes can be used as a basis for preventing the release of message contents and can be combined with other countermeasures.

Banks and insurance companies that routinely use certified DES equipment complained at NSA's initial announcement on DES certification. Their complaints were based on perceived security vulnerability.

The NSA decided DES was vulnerable. The DES algorithm is still good and has not been broken. The NSA withdrew its support because the algorithm had been around for more than ten years. Potential adversaries have had this time to work at breaking the code. Because the DES algorithm protects many valuable resources, such as banks and the federal government's, an adversary has the motive to break it.

In order to analyze the reaction to the NSA's decision, consider how most commercial organizations use DES. Businesses compare their risk of loss from security vulnerability with the cost of improving their security systems. They are not going to spend more to fix a problem than they would lose if they do not fix it. This is the basis for risk analysis.

Businesses also compare potential losses against their insurance premiums. If a business can buy insurance to cover potential losses for less money than security equipment will cost, they will do just that.

Insurance companies assume this risk. They also try to reduce their losses from claims by offering reduced premiums to those businesses that install security equipment. Of the many algorithms available, only DES is approved by the U.S. government. Insurance companies cover certified products, so if businesses desire reduced premiums, they have to buy certified DES security equipment.

The NSA is suggesting no further spending on DES equipment, and that firms should replace it with new equipment and algorithms. The reaction from most companies to this suggestion was they did not want to spend more on security, and if the NSA stops certifying DES, insurance companies will increase premiums.

AUTHENTICATION MECHANISMS

Authentication involves the actual application of the procedures for verifying the identity of a participant, including users, terminals, and networks. Personal authentication includes the use of passwords, badges, keys, fingerprints, and voiceprints. Except for the use of encryption keys for authentication purposes, these techniques all rely on the transmission of a stream of identifying bits to the location site performing the authentication.

The authentication mechanisms then become dependent on communication security. Authentication, unlike encryption, does not protect against eavesdropping because all messages are sent in cleartext. Because encryption and authentication are similar, most systems can perform both functions, and the cost is about the same. There must be measures to prevent the release of authentication information and to maintain the integrity of the communication paths after authentication has been performed. The first requirement can be achieved with message contents countermeasures while the second requires message-stream-modification countermeasures.

AUTHENTICATION LIFETIME

The effective use of authentication techniques depends on the lifetime of the information transmitted during the authentication procedure. This lifetime can be based on the number of times the information is used or on a time interval over which the information is acceptable.

Fingerprints and passwords might have no set lifetime, while one-time badges, encryption, or physical keys might be given a set time value.

This characterization of authentication methods based on the lifetime of the information transmitted limits the damage that can result due to the exposure of the information.

When the authentication information is static and is left exposed, it allows an intruder the time to develop and test penetration schemes. Dynamic authentication information is much harder to penetrate and might not be useful to the intruder at all. This includes one-time password or encryption key schemes. Some of these can be chained in which the next password or key is transmitted as an epilogue.

Chained authentications do not provide the same security as the use of true limited-lifetime characteristics, but they are superior to static procedures.

In the network environment, centralized authentication services are needed. There are often accounts on multiple service hosts, and if the same authentication methods were used at each host, penetration of any one of these would allow access to the other hosts.

If dynamic authentication consisting of one term passwords is used, it is difficult to maintain the password lists on each of the hosts without some form of centralized control. Personal authentication is commonly used for both access control and accounting purposes. Central authentication simplifies procedures for those who need to access only a single host in the network because it reduces with the need for multiple or layered authentication. Encryption keys used commonly are for authentication purposes in these environments. The use of an encryption module in every terminal can create a high degree of security, but the cost of modifying all of the existing terminals can be high. If the users access the network via terminals attached to server hosts, an encrypted communication path from the network through the serve host and back to the terminal can be difficult to implement.

Primary keys can be powerful and inexpensive, but the distribution of keys at the devices can present some problems. The tenants of separation of privilege and least privilege suggest that key distribution and personal authentication services be independent because encryption keys are a common mechanism for personal authentication.

The solution is centralized key distribution. A secure authentication server host is needed to act as an intermediary, carrying out the user authentication procedures and forwarding the results to the service hosts.

Each authentication server host can have a specific area of responsibility over which it provides services to users and hosts. The authentication server hosts can be grouped to provide load sharing.

The different groups can be provided with boundaries with a trust function among the groups. Simple protocols can be used to handle authentication across the boundaries.

An authentication server can be used to store the authentication information. The authentication server might act as the user's agent in carrying out the authentication procedures. The authentication server then appears as a user to each service host. When passwords are used, instead of storing the password at the authentication server, the image of the password based upon a one-way transformation is stored. The transformation is then submitted as the password during the authentication procedure.

NETWORK-SECURING TECHNIQUES

The most widely used techniques for securing a network are encryption, authentication, and access control.

Encryption minimizes the chances of a break-in by making the data unintelligible to anyone who does not have the key to do the encrypting (coding) and decrypting (decoding). The code might substitute one piece of data for another, or it might use transposition, which changes the order of the data in some specific manner. In either case, the key, which can be contained on a microchip, provides the means for data to be encrypted on the sending end and decrypted on the receiving end.

Besides protecting transmission privacy, encryption can also detect a break-in because any alteration of the data will result in the receiver getting a garbled message.

One of the most important considerations when developing an encryption program is to follow an ANSI standard or another accepted standard. The best defense when being sued for a breach of security is that an accepted standard was followed.

In 1977, the U.S. Bureau of Standards issued its Data Encryption Standard (DES), which provides a fixed scrambling algorithm.

While encryption and authentication protect the data en route, they do not protect unauthorized use of nodes that might contain the key. To control access to these nodes requires some simple precautions. Idle terminals should be disconnected and the host programmed to refuse any transmission from them. Also, the host should cut off a terminal if too much time elapses without any input after it requests a user password.

Passwords should be changed frequently, and they should not be based on personal data that can be traced such as a birthday. Passwords should not include words commonly found in dictionaries because these can be broken using programs that modify spellchecking software.

A recent means of preventing unauthorized use employs biometric control devices that read users physical characteristics, such as fingerprints, and match them with user data from its database. Only users whose physical characteristics are authorized to gain access are allowed to log on.

Access and security problems and solutions often depend on the network topology. For example, in a star network in which all traffic goes through the host, only three nodes should ever handle any transmission: the sending node, the receiving node, and the hub or host. This reduces the potential points of attack. A star network requires strenuous security to protect the host because all transmissions go through it.

One method of protecting a star network is to use access-control software. These packages provide a means of verifying the access authority of those who try to use the system. The software intercepts users' requests for system access and verifies their passwords and identifications. The users' IDs determine their level of system access.

In a bus or hierarchical topology, more nodes can be involved in any transmission, widening the possibility of a security break.

Another factor to consider is whether a network is broadband or baseband. Broadband networks use more complex signals, and eavesdropping on them is more difficult. But an attacker who successfully breaks into a broadband network is usually rewarded with more information than one who gains access to a baseband net.

PLANNING FOR SECURITY

Generally, an authorized, technically competent insider represents the major security threat. Control over information must be intrinsic to the hardware or software. Security should not be something that is added on as an afterthought.

Some simple network-security controls are implemented on a single, nodal computer servicing a star cluster of dumb terminals. The management of distributed networks employing components from many suppliers is a more complex task.

More secure networks are possible using a security architecture that conforms to the philosophy of open systems. A network security architecture can make use of the following characteristics.

The operating systems in the users' computers can police the application. A part of the hardware and software can include a secure computer kernel. It could be implemented as microcode and control a user's access to network resources through adherence to centrally imposed rules.

Computer networks should be managed through a range of supportive services. Such services could reside in user nodes or in stand-alone servers and complemented by a certified series of cryptographic processes.

SECURITY BARRIERS

General-purpose, multi-user workstations should employ both physical and logical controls, including tamper resistance, anti-electronic eavesdropping features, and user recognition authentication and authorization, which could be based on both logical and physiological properties.

Security barriers can be graded by power and cost (FIG. 5-5). The barriers start with physical constraints at the simplest level and add administrative, logical, and encryption barriers for more complete security.

A physical barrier continues to be one of the most relevant security measures that can be applied in a distributed network. The security controls needed to manage a single, centralized computer center are generally well understood.

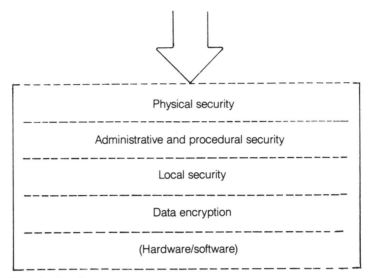

Fig. 5-5. Security barriers.

The modern workstation represents another problem. The very characteristics that make the workstations popular create a number of security weaknesses. For example, modular physical and logical construction allow the information and operating systems procedures to be removed easily, even when they are in hardware. An intruder can physically remove the data-storage device or alter confidential data by changing the properties of the systems.

Workstations with security features such as flat-screen technology, physical tamper-proofing, and electronic shielding are needed.

Control of workstation physical access using an identification token or magnetic stripped card is common. These smart cards with a built-in microprocessor and programmable RAM allow comprehensive identification and authentication. These are being used in a wide range of nonsecurity applications to hold data that can facilitate the transfer and control of user information.

Smart cards can be the first step in making the user responsible for data security. Another barrier is developing a set of security administration and support policies. Administrative barriers can range from corporate directives to the operational surveillance of staff entering secure facilities.

A security program can cover users of the services, the service suppliers (internal and external), and the auditing functions (financial, security, and quality assurance).

Increasing involvement in the technology of electronic information programs and in the operational involvement in distribution of network encryption keys would be required. An expanded role of the internal auditor with overall responsibility for ensuring the protection of all company assets could be perceived as a significant expansion over an earlier limited financial role. These wider administrative issues will need to be integrated in the detailed logical controls programmed into the various network services.

The rules of information and resource authorization, creation, and use should reflect the overall control needs of the organization. The user who can access the database and the telecommunications line that has the necessary control mechanisms to carry the classified message will need to be described with the correct privileges and control attributes.

Logical controls act as security barriers to ensure that only those authorized can access various networked resources. Most operating systems rely on externally administered access-control rules. The quality of control is primarily at the discretion of a human administrator at a centrally managed unit. Network-wide technology-driven security kernels with universally understood authorization policies are needed in operating-system implementations.

User identification and authentication generally is handled by a simple passwording system either built in or adjunct to the user application. Once a password is entered in a system, it is compared with a table entry. The data usually is one-way encoded (encrypted) to avoid tampering.

After the password is authenticated, the user can be authorized to access network resources based on privileges.

Public-key algorithms allow the encryption of messages by obtaining a unique logical key from a published source. The process cannot be reversed by using the same key. A personalized key known only to the recipient decodes the message. Such keys which are a long, complex string of numbers need to be well protected. The underlying security is based on the difficulty involved in factoring large numbers. The encryption keys are being packaged as portable tokens or smart cards.

Private-key systems allow both the encryption and decryption of messages by issuing common keys that transform data according to agreed upon algorithms. Encryption processes can be handled with very large-scale integration (VLSI) chips.

The use of VLSI chips reduces the overhead of data encoding and decoding. VLSI implementations can provide an inexpensive means to secure broadcast-based local-area network services in modern multiple-occupancy buildings.

STANDARD EVOLUTION

Government agencies have invested heavily in encipherment systems for ensuring privacy of the flow of classified data internally as well as across communications networks. The role of the National Security Agency in secure, encrypted networks to increasingly define and control the use of encryption technology also applies to civilian networks.

The use of a long-term replacement for the Data Encryption Standard will affect all those currently developing secure network products. The Department of Defense's Trusted Computer Systems Evaluation Criteria, known as the "Orange Book," covers the central/nodal computer operating system. It can be augmented by other DOD color books. This could include a Gray Book covering trusted databases and a Brown Book, which would become the government standard for trusted networks. The colors selected could become de facto standards that can be used to evaluate security features.

The Trusted Computer Systems Criteria Standard is being used to certify operating systems products. Products that conform to an Orange Book high-C or low-B rating are considered as appropriate for nondefense commercial nodes and distributed networks. Commercial computers that conform to the B level can apply rules from a trusted part of the computer, a trusted computer database, or a kernel that is incapable of access during operational running.

When a computer operating system conforms to the B level, the controls pertaining to sensitive data or important computer or network services cannot be changed at the discretion of the user. B-level security implemented for commercial purposes will normally be supplemented by a set of discretionary security features.

The C-level classification can provide commercial security features with a set of user-driven rules. In terms of network-wide controls, C-level security is only as good as the sum of the individual user capabilities. The C-level classification provides minimal security features at the network level beyond interfacing with basic network mechanisms such as encryption.

The security criteria in the color books, in combination with the network architecture, allows products and services to have a common direction to the level of security necessary within various types of network solutions. FIGURE 5-6 illustrates how a user might decide on the level of security for a network node. Productive and supportive services occupy different configurations of the operating environment with the necessary DOD Orange Book security classification. Multiple supportive service implementations or mixed productive – supportive services in the end-system node imply controls at the B level.

Supportive services used by productive services include authentication, directories and routing tables, encipherment and encryption services, auditing, service accounting, and network-management software.

The interactions in a secure network are outlined below. A user will interface with a service (a user sponsor in a local workstation). The user sponsor calls upon services distributed through the network. The facilities can reside on different computers under the control of different operating systems with various security levels and functionality. A common security policy through the network is an administrative objective.

The necessary securely held privileges will be dependent upon responsibility, status, seniority, need to know, or other criteria. The privileges will be

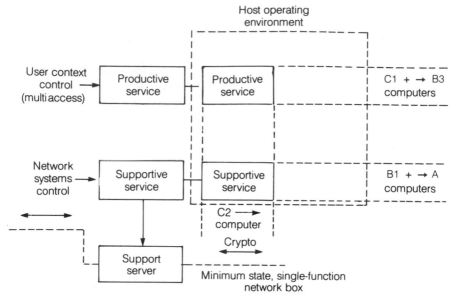

Fig. 5-6. Network node security levels for productive and supportive services housed in different configurations.

stored and transmitted encrypted to avoid changes by those tapping into the network.

The authorized network user should be able to travel freely throughout the network, through private local network domains and public networks. Prompting at regular intervals can prevent human compromise until the desired computer or network resources are reached. Secure-authorization network servers, some of which are at network gateways, can be policed by security policies that compare control privileges with access lists and associated controls. A combination rule-based and identity-based universal policy can result in logical access authorization. The unauthorized should be unable to compromise the data or amend the control processes no matter how technically competent they are.

The data-security officer has to know what controls will work where and which ones can be replicated on different devices. The security built into applications also must be considered in this context. If mainframe has RACF, ACF2, or Top Secret running on it, the controls in that package could be preferable to the security capabilities built into the production applications.

MULTIVENDOR SECURITY

In a multivendor network in which there are several security domains, each has a different span of control and a different degree and method of protection.

When data are moved from one security domain to another, they lose the protection afforded by the first domain unless the new one has some mechanisms for indicators of data sensitivity or implementing similar controls. Data moving within such a network might be compromised in a relatively unprotected domain. Recognition of the relative strength of security domains can aid decisions on how to route sensitive data through the network.

Sensitivity indicators also should be part of the data traversing the network so that available controls can be utilized. These indicators must be added in the native domain by the application before the data enters the network. This might require modifying the application so it can identify and tag sensitive data.

A common scenario to a prospective user in a multivendor network environment is a series of menu screens, several of which might require unique IDs and passwords. Before getting to any productive work, a user might have to supply three or four separate IDs in the correct sequence and tied to the current password. Live IDs and passwords of ten can be found stuck to a number of terminals.

IMPLEMENTATION GUIDELINES

Application-support groups usually entertain security's request for common IDs and passwords as long as it is the other applications that must change their code and standards. Sometimes an application cannot be modified to accommodate a generalized password scheme because of contracts that stipulate that any changes to system code invalidate maintenance agreements.

If security controls interfere with the ability of users to do their job, they will find a way around those controls. Even if a unified ID/password standard can be agreed upon, installation of such a system will require several interface mechanisms to allow migration between applications. The long-term benefits of such a system far outweigh all of these initial drawbacks.

Burdensome layers of control can make a system user antagonistic. User tolerance can be improved a number of ways.

Educate the user with a security-awareness program. Users should be told why security is necessary and why its implementation uses the methods chosen.

Be realistic about complex sign-ons, IDs, and passwords. Although ten or more digits of randomly generated, alphanumeric passwords are secure, humans will not remember them. Security vanishes when IDs and passwords are written on pads or terminals. Passwords should be changed every 30 to 90 days. If a password never changes, it might as well be published or eliminated.

Get application groups to communicate with representatives of each involved department. Their support needs to be gained so users will be more likely to adhere to the guidelines that are developed.

Publish security error messages to clarify the environment, publish the various security-related error messages put out by each component of the system. Include an explanation of where each message comes from and what it means.

Users know what their job requirements are and what needs to be protected. If the constraints and the basic mechanics for implementing controls are known, users can give relevant input for maintaining adequate security without strangling productivity.

Adding a simple control to close a minor exposure can have disastrous effects on downstream productivity. Before a control is implemented, its effects should be researched and understood throughout the network, not just in the domain of installation.

Data proteciton must be considered with any type of network. Direct-wired networks such as local-area implementations present special problems. Terminals as well as cabling must be protected. Cables should be kept in secure areas. Cables can be sheathed in a heavy metal casting to deter intrusion.

Sheaths with a positive gas pressure are also available. The pressure is monitored from a centralized location. If the sheath is broken, the pressure drops and an alarm is issued. Shielding also is necessary for listening devices that can pick up signals from cables.

An audit trail should be in place so if a break does occur, the activity on any node can be documented and the progress of the transaction can be traced. One way to keep an audit trail is to have the computer keep a snapshot of each transaction. The snapshot should contain the date and time the transaction was entered or delivered; a terminal and system sequence number; originating or destination department; operator ID (but not the password); message length; and a description of the message's route.

Snapshots should be kept for a proper length of time. This time period depends on the data sensitivity, the relative difficulty in reconstructing the message, and the probability that security procedures will discover an attack quickly.

No matter how sensible a particular set of controls seems, it might not be acceptable in a specific environment. Alienating department managers, network technicians or users does little to enhance the reputation of the auditor and can lower the chance for success of the network security program.

A complex multivendor network is a tool to help the organization achieve its objectives, usually measured in terms of market share. Security is not an end in itself.

Security organizations are chartered to protect networks, programs, data and the associated processing capability required to assure the continuity of the business.

Security programs must comprehend the objectives of the organization and focus on protecting the elements most crucial to the achievement of these objectives.

6

The OSI Model

In this chapter we will consider the protocols used in the Open Systems Interconnection (OSI) reference model.

INTRODUCTION

The evolution of local-area standards began with EtherNet and evolved to the current IEEE 802 standards, which cover several types of local-area configurations. Network protocols support a variety of combinations of topology, transmission medium, signal-encoding technique, and data rate. Protocols and standards are involved with more than just the protocols internal to the network. Standards for interfacing the terminals and computers to the network are also necessary. There are also the protocols for network management and control operations.

The key to the development of the local-area network market is the availability of a low-cost interface. The cost of connecting equipment to a LAN must be much less than the cost of the equipment alone. This requirement, plus the complexity of the LAN protocols, dictates a very large-scale integration solution. However, chip manufacturers are reluctant to commit the necessary resources unless there is a high-volume market. A LAN standard would ensure that volume and also allow equipment from a variety of manufacturers to intercommunicate.

This is the rationale of IEEE 802. Its standards are being processed by the International Organization for Standardization (ISO). The standards are in the form of a three-layer communications architecture, which is matched against the ISO's Open Systems Interconnection (OSI) reference model in FIG. 6-1.

The logical link control (LLC) layer provides for the exchange of data between service access points (SAPs), which are multiplexed over a single physical connection to the LAN. The LLC provides for both a connectionless,

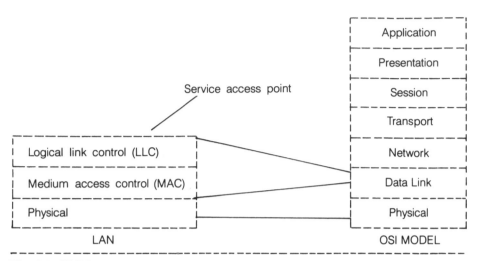

Fig. 6-1. IEEE 802 layers.

datagram type and a connection-oriented, virtual-circuit-type service. Both the protocol and the frame format resemble those of the high-level data-link control (HDLC).

PROTOCOLS AND THE OSI MODEL

In terms of the OSI reference model, the protocols internal to a local network occupy the two lower layers: the physical layer and the data-link layer. The physical layer protocol covers the mechanical, electrical, functional, and procedural functions. RS-232, for example, defines a standard 25-pin connector, specific electrical signal levels assigned to functional pins, and the procedures for initializing, sending, and receiving the data.

IEEE 802

The IEEE began project 802 for local network standards in 1980. The general charter of the project covers the two lower layers of the OSI reference model and the interface to the network layer. The data-link layer is divided into two sublayers as shown in FIG. 6-1.

An access-unit interface takes the place of the transceiver cable that is used in EtherNet. It is defined between the component containing both the link layer and physical signaling and the physical medium attachment to the transmission media.

The IEEE 802 standard is actually a family of standards as shown in TABLE 6-1.

Table 6-1. IEEE 802 Family.

	IEEE 802.2 Logical link control		
Link layer			
Physical layer	IEEE 802.3 CSMA/CD Bus	802.4 Token Bus	802.5 Token Ring

IEEE 802.2 covers the logical link-control sublayer. The three main options involving the MAC sublayer and the physical layer allow:

- CSMA/CD access on a bus topology
- Token access on a bus topology
- Token protocol on a ring topology

FIGURE 6-1 shows that the LAN communication architecture involves three layers: physical, MAC, and LLC. The physical layer provides for attachment to the medium. The MAC enables multiple devices to share the medium's capacity in an orderly fashion. The LLC provides for data-link control across the network.

Higher layers of software will make use of the LLC. This includes the Network and Transport layers in the OSI model. Above the transport layer are user-oriented layers (session, presentation, application) that handle communication functions.

The data-link-layer protocol provides functions such as error and access control on a logical data link. Many local network protocols resemble HDLC at the data-link layer. HDLC uses a cyclic redundancy checksum (CRC) for error detection and polling to control which station has access to a multipoint link. Local network access-control schemes include carrier-sense multiple access with collision detection (CSMA/CD) and token-passing schemes that are related to polling. The OSI reference model for point-to-point lines is shown in TABLE 6-2.

The medium access control (MAC) layer provides for the regulation of access to the shared LAN transmission medium. For each MAC specification (carrier-sense multiple access with collision detection and token ring), a physical layer specification tailored to the topology and MAC algorithm of the corresponding MAC protocol is provided.

Table 6-2. OSI Model for Local Networks.

OSI Reference	Protocol for Point-to-Point	Control for Local-Area Network
Physical	RS-232	Baseband or broadband signals
Data link	HDLC	CSMA/CD or token passing

Table 6-3. IEEE 82 Options.

CSMA/CD	
Baseband 50-ohm coax	Broadband 75-ohm coax
Manchester coding	
10 Mb/s	2, 10 Mb/s
Token Bus	
Single-channel 75-ohm coax	Broadband 75-ohm coax
Continuous and coherent FSK	Multilevel, duobinary AM/PSK
1, 5, 10 Mb/s	1, 5, 10 Mb/s
Token Ring	
Baseband 150-ohm twisted-pair	Baseband 75-ohm coax
Differential Manchester	
1, 4 Mb/s	4, 20, 40 Mb/s

The IEEE-802 standards are designed for low-to-medium-speed LANs of up to 20 Mbps. Higher speed LANs are within the ANSI X3T9 technical committee's work on I/O interface standards.

A subcommittee known as X379.5 has developed a proposed standard called the Fiber Distributed Data Interface (FDDI), which is to be adopted as an ANSI standard. FDDI specifies the MAC and physical layers and assumes the use of the IEEE-082 LLC layer. The MAC layer is a token-ring specification, similar to that of IEEE 802. The physical layer specifies a 100-Mbps optical-fiber ring.

Along with the three major options, there are a number of suboptions for different transmission media, signaling methods, and data rates. TABLE 6-3 summarizes these options.

FIGURE 6-1 also indicates how the higher layers of software make use of the LLC. The LLC provides a service of transmitting frames of data across the network, and that service is invoked by a higher layer of software. The IEEE 802 standards use this layered architecture.

Because the three LAN layers correspond to the lowest two layers (physical and data link) of the ISO's Open Systems Interconnection (OSI) model, the next layer would then be the network layer. In a LAN, most functions associated with the network layer are not required.

The LLC layer provides for the needed routing and addressing to transmit protocol data units from source to destination. If a connection-oriented LLC service is used, the LLC layer can provide the logical connections, flow control, and error control.

The ISO connectionless, and network-layer standard provides an internetworking protocol and is needed so the hosts on the LAN can be connected via a gateway to hosts on other subnetworks.*

---

*See "Internetworking in an OSI Environment," *Data Communications*, May, 1986, p. 118; and also "Of Local Networks, Protocols, and the OSI Reference Model," *Data Communications*, November, 1984, p. 129.

The LLC layer moves frames of data from one station on the LAN to another. The network layer provides internetworking capability. The transport layer provides end-to-end reliability.

As shown in FIG. 6-2, the transport entity encapsulates the data with a transport header and passes the resulting unit to the network-layer entity, which adds its own header and passes the resulting unit to the LLC. The LLC in turn adds its own header and passes the resulting unit to the MAC.

The MAC produces a frame with both a header and a trailer, and this frame is transmitted across the LAN. The MAC frame includes a destination station address. The frame is copied by the station with the address. The user's block of data then moves up through the layers, with the appropriate headers and trailers stripped off at each layer.

As discussed above, there are two types of operation possible using the logical link-control procedures. Type 1 refers to a connectionless service and type 2 to a connection-oriented service. Type 1 is like a datagram because there is no need for the establishment of a logical data link. There is no acknowledgment, flow control, or error recovery. Either individual or group addressing can be used. In type-2 operation, a logical data link is established, acknowledgments are provided, and only individual addressing is allowed.

There are also two classes of logical link control. Class I is allowed for type-1 operation only, but class II can be used for both type-1 and type-2 operations. All stations can support both connectionless and connection-oriented operations.

The logical link-control sublayer includes the link-control procedures and interface service specifications to the network layer and the medium-access-control sublayer.

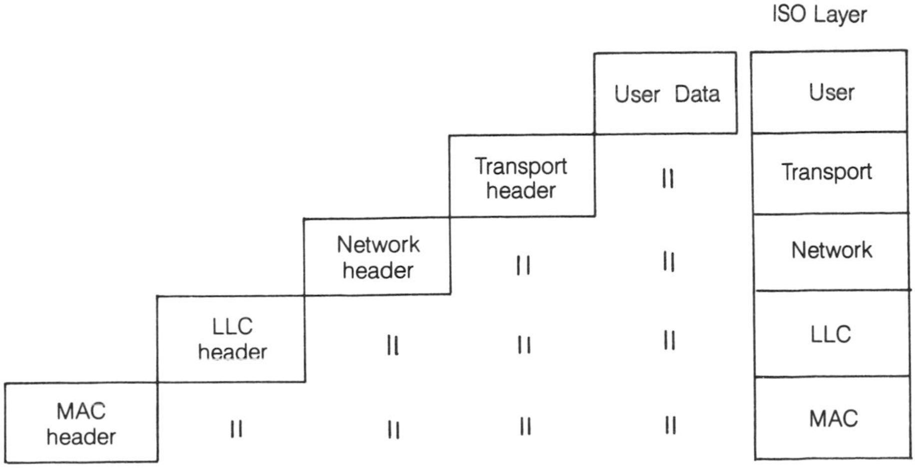

ISO Layer

				User Data	User
			Transport header	II	Transport
		Network header	II	II	Network
	LLC header	II	II	II	LLC
MAC header	II	II	II	II	MAC

Fig. 6-2. As data moves up or down through the layers, the headers are stripped off or added.

The interface-service specification to the network layer defines the calls for an unacknowledged connectionless service. This is a datagram type of service in that the stations exchange data units without the establishment of a data-link connection or acknowledgments. A connection-oriented service can also be used as an option.

The logical link-control procedures have evolved from the asynchronous balanced mode used in HDLC and ADCCP (ANSI X3.66). These have been extended for the multistation, multiaccess environment of LANs.

In the point-to-point or multipoint configurations supported by HDLC/ADCCP all transmissions are to and from a primary station. In a local network, a logical data link can exist between any pair of stations. A single station can have multiple data exchanges simultaneously with several different stations. This type of exchange is treated via a port address to a higher-level protocol within the station and will be discussed shortly.

The format of the logical link-control protocol data unit is shown below:

DEST SAP	SOURCE SAP	CONTROL	INFO
1 OCTET (8 bits)			

The destination service-access point (SAP) and source service-access point represent individual addresses. DEST SAP might also be used as a group address.

The control field is used for the commands and responses. Connectionless data exchanges use an unnumbered information (UI) command. Connection-oriented data exchanges use an information (I) command and response, along with receive ready (RR), reject (REJ), and receive not ready (RNR) commands and responses.

FRONT-END PROCESSING

The discussion thus far assumes that all layers of the architecture are implemented in the same processor. Some of the lower layers might be offloaded onto a front-end processor (FEP). A network can consist of not only the transmission medium, but also a set of intelligent devices that implement the network protocols. These devices also provide an interface for attaching the devices (terminals or computers.) These intelligent devices are the FEPs.

The FEP architecture is common in independent-vendor local networks. The vendors sell networks, but not data-processing equipment.

A serial-communications link (RS-232) is used between the front-end processor and the attached devices. A physical-layer protocol and a link-layer protocol are needed to exchange data across the link. A burst-to-front-end (HFP) is also needed.

Table 6-4. Logical-Link-Control-Service Primitives.

Unacknowledged Connectionless Service

L.DATA.request (local-address,remote-address,1-sdu,service-class)

L-DATA.indication (local-address,remote-address,1-sdu,service-class)

Acknowledged Connectionless Service

L-DATA-ACK.request (local-address,remote-address,1-sdu,service-class)

L-DATA-ACK.indication (local-address,remote-address,1-sdu,service-class)

L-DATA-ACK-STATUS.indication (local-address,remote-address,service-class,status)

L-REPLY.request (local-address,remote-address,1-sdu,service-class)

L-REPLY,indication (local-address,remote-address,1-sdu,service-class)

L-REPLY-STATUS,indication local-address,remote-address,1-sdu,service-class,status)

L-REPLY-UPDATE.request (local-address,1-sdu)

L-REPLY-UPDATE-STATUS.indication (local-address,status)

Connection-Oriented Service

L-DATA-CONNECT.request (local-address,remote-address,1-sdu)

L-DATA-CONNECT.indication (local-address,remote-address,1-sdu)

L-DATA-CONNECT.confirm (local-address,remote-address,status)

L-CONNECT.request (local-address,remote-address,service-class)

L-CONNECT.indication (local-address,remote-address,status,service-class)

L-CONNECT.confirm (local-address,remote-address,status,service-class)

L-DISCONNECT.request (local-address,remote-address)

L-DISCONNECT.indication (local-address,remote address,reason)

L-DISCONNECT.confirm (local-address,remote-address,status)

L-RESET.request (local-address,remote-address)

L-RESET.indication (local-address,remote-address,reason)

L-RESET.confirm (local-address,remote-address,status)

L-CONNECTION-FLOWCONTROL.request (local-address,remote-address,amount)

L-CONNECTION-FLOWCONTROL.indication (local-address,remote-address,amount)

For a front-end whose highest layer is LLC, it is called an LLC-HFP. In an integrated architecture, the network entity passes a block of data to the LLC for transmission. The LLC commands for this data block are contained in the link-control primitives used by the network layer to invoke the LLC. These link-control primitives are shown in TABLE 6-4. For the case of unacknowledged connectionless service, the network entity would use the following primitives and parameters to invoke the LLC.

L-DATA request (local-address, remote address, 1-sdu, service-class) where:

Local address	=	local LLC service access point
Remote address	=	destination station (MAC) address and destination LLC service access point
1-sdu	=	block of data being passed to LLC for transmission
Service class	=	desired priority

If the LLC is invoked by a subroutine call, the L-DATA request call would be a machine-language subroutine. The implementation of this interface depends on the machine language and operating system utilized.

The subroutine-call approach can be used when the network and LLC entities execute in the same processor. In the case when they run on separate processors, however, a way is needed for the network and LLC entities to exchange primitives and parameters. This is the function of the LLC-HFP. It provides a way for the network and the LLC entities to communicate.

In the host, the LLC-HFP entity presents an interface with the network entity that emuluates the LLC. This allows the network entity to use calls such as L-DATA request as if the LLC and network were running on the same processor.

The network entity passes the user's data plus a network header to LLC-HFP, using the L-DATA request call. LLC-HFP appends a header and passes the result to the link layer. The link layer is a point-to-point data link-control protocol, such as:

- High-level data link control (HDLC)
- The header identifies
- The user invoking the services of the LLC
- The primitive being invoked
- The parameters associated with that primitive

This case assumes the data-link protocol, such as HDLC, is present across the host-FEP interface. If the FEP is a communications board, then there will be no link and physical-layer protocols. In many cases, the FEP and host processor are connected to the same bus and exchange information through a common main memory. The FEP typically uses DMA. A specific area of memory is shared and used to construct a system control block for communications. This system control block fulfills the same function as a header.

A typical block format is shown below for the first three words (TABLE 6-5).

Whether the host-FEP exchange is achieved by a header or by a system control block, the specification of that exchange is needed. The specification

Table 6-5. Block Format.

Type	Word	Description
Status	1	Network interface unit (NU) or host status information
Directive	2	Management-related commands and acknowledgments
Command Pointer	3	Points to an area of memory with one or more commands Each command consists of a command code and parameter list

Table 6-6. ISO Transport-Service Primitives.

T-CONNECT.request (Called Address,Calling Address,Expedited Data Option,Quality of Service,Data)

T-CONNECT.indication (Called Address,Calling Address,Expedited Data Option,Quality of Service,Data)

T-CONNECT.response (Quality of Service,Responding Address,Expedited Data Option,Data)

T-CONNECT.confirm (Quality of Service,Responding Address,Expedited Data Option, Data)

T-DISCONNECT.request (Data)

T-DISCONNECT.indication (Disconnect Reason,Data)

T-DATA.request (Data)

T-DATA.indication (Data)

T-EXPEDITED-DATA.request (Data)

T-EXPEDITED-DATA.indication (Data)

---

should include not only the formats, but also functions such as initialization flow control, management, and recovery. When both the FEP and the host are provided by the same vendor, the specification can be proprietary. If the FEP and the host are from different vendors, then a standard specification can be used.

TABLE 6-6 lists the primitives for the ISO transport protocol standard. The ISO transport protocol standard, like the LLC, uses the concept of a service access point (SAP), so a transport entity can have multiple users.

The first six primitives in the table are used to establish and remove the logical connections between transport SAPs. The quality-of-service parameter allows the transport user to request specific transmission services, such as priority and security levels. The remaining primitives are concerned with data transfers.

The exact functions performed by the HFP depends on the architecture employed. An LLC-HFP needs to pass different primitives and parameters than a TP-HFP.

Basically, the HFP must examine each command that it receives, verify the parameters associated with the command, and reformat as necessary for communication. Other characteristics of an HFP must be connection-oriented in order to maintain the state of a dialogue with each user of the HFP. It also must be multiplexed to allow multiple higher level users as well as individual flow control by the user. The characteristics of several local networks are shown in TABLE 6-7.

Table 6-7. Local-Network Characteristics.

Network	Standard	Media	Modulation	Access	Topology
EtherNet	IEEE 802.3	Coax/Fiber	Baseband	Contention	Bus
GM MAP	IEEE 802.4	Coax/Fiber	Broadband	Token passing	Bus
	IEEE 802.5	Twisted-Pair Coax/Fiber	Baseband	Token passing	Ring
	IEEE 802.6	Coax/Fiber	Baseband	Time slot	Ring
	ANSI X3T9	Fiber	Baseband	Token passing	Ring
PCnet	—	Coax	Broadband	Contention	Bus
ARCnet	—	Coax	Baseband	Token passing	Bus

ACCESS METHODS

Access methods stem from the distributed network intelligence. One method used to access the network medium is through contention. Here any unit on the network begins to send data whenever it wants.

Because two devices sending signals concurrently will garble the data, most contention schemes require that a station first listen to the network and then send data only when it is idle.

Additional problems can arise if two stations begin transmitting at the same time, which results in a collision of the data on the network.

The CSMA/CD access protocol is one of the options in IEEE 802. This option relates to the ISO layers as shown in FIG. 6-3. The physical medium attachment unit (MAU) has baseband and broadband options.

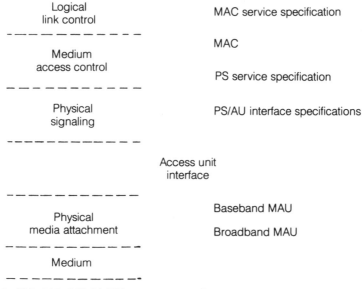

Fig. 6-3. ISO 802 CSMA/CD access protocol.

Collision detection is implemented to reduce the time that collisions exist in a network and improve the overall network throughout. Contention systems function with error checking, which result is garbled messages being repeated.

Another access method is token passing. A special message (the token) is sent from one station to another network. The token is used to grant permission to transmit a message on the network. A unit can transmit only after being granted permission by receiving a token indicating that the network is idle.

Problems can arise in passing the token around the network. A station can go off-line and be unavailable to receive the token and pass it on. The token can be distorted by noise so that it is never detected as being received. The network usually has one unit to serve as a watchdog or controller to handle these conditions.

The time-slot type of access operates by allowing each unit to send data only during a designated interval of time. No single station can use the full capacity of the network because each unit can transmit only during a small percentage of the total time that is available on the network. The effective use of the communication system can suffer with this method because the capacity required is not uniform from station to station.

In the carrier-sense multiple access with collision detection (CSMA/CD) protocol, the MAC sublayer constructs the frame from the data supplied by the logical link control. A carrier-sense signal at the physical layer then is monitored for traffic, and if the response is positive, transmission is deferred. If the response is negative or clear, a frame transmission is initiated after a fixed delay (9.6 microseconds for a 10-Mbps system) in order to provide recovery time for other stations.

The medium is monitored by the physical layer for collisions, and a collision-detect signal is sent to the MAC sublayer in the event of a collision.* Transmission is terminated, and a retransmission is attempted after a random delay.

If repeated collisions occur, the retransmission is repeated, but the random delay is doubled with each attempt. This is called *binary exponential backoff*.

The delay interval is doubled for each of the first ten retransmissions, and then stays the same up to the 16th retransmission. The retransmission is abandoned after 16 tries and an error condition is signaled.

A collision window time is allowed during the initial part of transmission before the transmitted signal has had time to propagate to all stations and a return signal propagate back. The time to acquire the medium is thus based on the round-trip propagation time of the physical layer. After the collision window has passed, the station will have acquired the medium and subsequent collisions will be avoided.

The retransmission delay is a function of the slot time, which exceeds the round-trip propagation time and the jam time. In a 10-Mbps baseband system,

---

*Collision handling begins with a bit sequence known as a *jam*. This sequence is 32 bits for a 10-Mbps system, and it is transmitted to ensure that the duration of the collision is sufficient to be noticed by the other transmitting stations.

the slot time is equal to 512 bit times. The backoff delay is an integral number of slot times.

A faster CSMA/CD network will out-perform a slow token-ring network up to a certain point. This point depends on the number of workstations on the network. When heavily loaded, a collision-sensing network degrades rapidly because of the exponential nature of the binary back-off algorithm employed. A token-ring network continues to degrade in an almost linear fashion as workstations are added and/or network traffic increases.

While it is generally not possible to calculate the response when using a collision-sensing protocol, it is a simple matter to determine token-ring response when the network traffic is known.

When a token-ring file-server network reaches a certain point with all nodes constantly transmitting maximum length packets, the response time can be determined by an analysis of disk access time on the file server. The network response is solely a factor of disk access when this point is reached in these types of networks.

The disk-access analysis indicates that response time increases in an exponential nature as workstations are added to a CSMA/CD network, while it increases in an almost linear fashion as workstations are added to a token-ring network.

A "faster" collision-sensing network will outperform a slower token ring up to a point. This depends on the number of transmitting nodes on the network and occurs in the graph where the two curves cross (see FIG. 6-4).

A 4-Mbyte/second token ring will perform as well as a similarly resourced 10-Mbyte/second CSMA/CD network (such as EtherNet) when 25 to 30 workstations are used. Beyond this point, the token ring should increasingly outperform the CSMA/CD network.

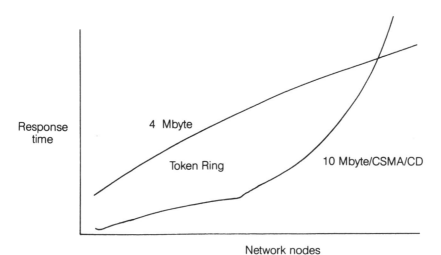

Fig. 6-4. Token Ring vs. CSMA/CD.

ETHERNET

In 1980, the EtherNet specification was released by Digital Equipment Corporation (DEC), Intel, and Xerox. The 1980 EtherNet specification was concerned with the two lower protocol layers of the ISO Model.

The EtherNet specification was based on the concept of an open system that allowed interoperability among different vendor's equipment.

EtherNet controller boards appeared in 1982, and EtherNet controller chips became available later that year.

At the physical layer, coaxial cable is used that normally does not exceed a few kilometers in length. This topology, known as *dendritic*, is not unlike the branching of a tree. Baseband signaling is employed at a data rate of 10 million bits per second.

At the link-control layer, the carrier-sense multiple access with collision detection (CSMA/CD) technique is used. This layer also defines a frame structure for messages similar to that used in HDLC.

The minimal configuration is a single segment of coaxial cable of up to 500 meters. The user devices are connected to the main cable using an access or transceiver cable that can be up to 50 meters in length. Most of the EtherNet protocol logic is implemented in the user device.

The access cable is connected to the main cable with a transceiver that performs the signal transmission and reception. Propagation delays are kept short so that the transceiver can detect collisions by comparing the outgoing and incoming signals. Transmission is stopped when collisions are detected.

A local network of more than 500 meters will use multiple segments that are connected with repeaters. A repeater operates at the bit level, and hence, at the physical layer of the OSI model. It is used to extend a particular section past the normal physical boundaries imposed by its electrical characteristics and the media used. The repeater restores voltage levels, sharpens signals, and sends them on their way to an additional section on the network. The repeater retransmits every message with a small delay that acts to cancel out the distance limitation of a single segment with the baseband signaling employed.

In a large building, one segment could be vertical, up a utility shaft, with segments spread out for each floor. A single segment can contain up to 100 stations.

A point-to-point link is used to connect segments that are some distance apart. This can be up to 1000 meters in length and acts as a repeater divided into two sections. The interconnected segments appear to be in one logical space to the CSMA/CD protocol. A maximum of two repeaters can be in the path between any two stations.

EtherNet is one of the options within the IEEE 802 local-network standards. There are some modifications from the 1980 specification, which has since been revised.

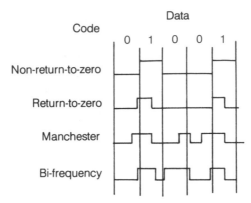

Fig. 6-5. Digital encoding techniques.

Although baseband allows simpler circuitry, it is limited in its information-carrying capacity. This capacity might meet the users' needs.

Broadband modulation requires rf circuitry (an rf modem) similar to CATV devices. It can deliver signals with various frequency bands, or channels. Each channel can carry a considerable amount of information, approaching the capacity of a baseband system. Optical systems rely on amplitude modulating a light beam, which is a form of carrier as in rf systems. These systems generally are considered to use baseband signaling, with the optical intensity variation similar to the voltage variation used in electrical baseband systems.

Related to modulation is digital encoding. The digital data must be represented by some form of code on the network. A simple code is *NDZ*,* in which each bit is represented by one of two states of the modulated signal. Other codes can be used to impart special characteristics to the signal in order to provide enhanced capabilities to the network hardware.

Manchester code is a type of biphase code. It employs two states to encode each bit of data, which doubles the signaling rate, or baud rate, on the network. This code is used in EtherNet applications and many fiberoptic links because they provide an average 50 percent duty cycle to the network signals. Common coding techniques are shown in FIG. 6-5.

Manchester coding is used for the transmission of data through the AUI. This type of coding combines the data and clock into a single bit stream as shown in FIG. 6-5.

A transition to the inverse state always occurs in the middle of each bit period, while the signal state during the first half of the bit period indicates the data value.

---

*Non return-to-zero.

ISO AND COS

Many organizations avoid depending on a single vendor by using a variety of computer hardware. In theory, this approach allows users to select whatever vendor's equipment is most appropriate for a given application. This multivendor tactic often leads to compatibility problems.

A major response to users' need for standards to support multivendor connectivity began in March 1977, when the International Standards Organization (ISO) started the task of building user and vendor consensus for its proposed Open Systems Interconnect (OSI) architecture and protocols. OSI can provide a means to open vendor-specific closed systems by linking them to each other via a common intermediate communications architecture. As a communications standard, OSI is not like a universal language that allows people to communicate. Instead of learning every language available, one would only need to learn one common language.

OSI's standard, layered stack of protocols can be implemented in parallel with a vendor's proprietary protocol stack.

OSI provides architectural flexibility through multiple protocol options at each of its seven layers. Application-specific subsets of OSI can be built by selecting a specific protocol option at each layer.

General Motors used this approach in its OSI-based Manufacturing Automation Protocol (MAP) architecture for factory floor applications. Boeing also used this approach to develop the Technical and Office Protocol architecture for technical office applications. Both companies seek to have their own sectioned OSI architectures accepted as standard within their specific industries.

The Corporation for Open System (COS) is a consortium of users and vendors with the task of monitoring and testing OSI protocol standards and controlling the development of OSI products.

The formation of COS was an important step toward connectivity. COS is dominated by vendors, with only 17 user companies among its 65 members. Also, user organizations such as the International Communications Association (ICA) and the Tele-Communications Association are not entitled to full membership in COS, and they must take an advisory role.

The large part played by vendors in COS has raised user concerns that computer makers will exploit the organization by promoting proprietary protocols, leaving users again with incompatible products.

Vendors have been participating in projects like the NBS OSInet, using their own software implementation of the International Standards Organization's OSI protocols. OSInet is an NBS sponsored program designed to promote open testing of OSI software and to speed the migration of government networks to OSI. Of OSInet's 22 members, IBM, AT&T, Digital Equipment Corp., General Motors, Hewlett-Packard, Honeywell-Bull, International Computer Ltd., Mitre, NCR Comten, Retix, Unisys, and Wang Laboratories, Inc., participated in this phase of OSI testing.

These test are a step toward bringing OSI-based products to market and assuring users that products can interoperate. Test participants see interoperability testing as a step toward bringing OSI products to market. This testing is not a substitute for conformance testing that ensures that a vendor's OSI implementation fully adheres to OSI standards.

In interoperability tests, OSInet participants test the protocols for the seven-layer OSI model and application corresponding to layers.

Users need communications software that incorporate all seven layers of the OSI model, and they want to see tests that check all those layers for multivendor interoperability.

Each of the OSInet participant's communications software incorporated the MAP 2.1 protocol suite, which includes: the ISO's File Transfer, Access, and Management (FTAM) application layer protocol; the full Basic Control Subset session-layer protocol; the Transport-Class 4 protocol; the Connectionless Network protocol; and the X.25 protocol.

OSInet used two local-area networks and one X.25 wide-area network. Vendors accessed the OSInet backbone network, which is based on AT&T's Accunet and Wang's WangPac transmission services. An X.25 gateway at NBS routed traffic to either the OSInet 802.3 or 802.4 local-area network. The vendor either sent or accessed a file on a host or microcomputer at NBS or at another vendor's location.

The OSInet participants did not test Layer 6, the session layer of the OSI model, because the protocols were not completed when testing began. That also was true for the Association Control Service Element, which is a protocol common to all layer 7 applications.

The next step for the OSInet interoperability project is to implement tests for all seven layers of OSI protocols and to test other application protocols, such as the X.400 electronic-messaging standard.

The Corporation for Open Systems (COS) also is planning to implement interoperability and conformance testing for OSI protocol implementation. COS is a consortium of both computer and communications vendors and users.

Both NBS and COS are doing similar work, although the primary purpose of NBS is to develop open systems for government users, while COS is driven by the needs of users in the private sector.

OSI CONNECTIVITY CASE STUDIES: IBM

During the late '70s, IBM clearly impeded competitors' efforts to design products that attached to Systems Network Architecture (SNA) networks. Few IBM architectures were open, and other vendors discovered many obstacles when they tried to design SNA products.

IBM mainframe users bought IBM products because they knew they could attach them to SNA networks. IBM views other suppliers' products as foreign and promoted pure SNA networks.

IBM's more recent networking strategy is to encourage other vendors to connect to SNA networks and develop products that conform to international standards. Much of this change is based on IBM's positive experience with its first completely open device, the Personal Computer.

IBM then encouraged third parties to design product enhancements. This was a major factor in establishing the Personal Computer as a de facto standard.

It is difficult today to be open unless you are also open to international standards. OSI and ISDN are becoming part of that definition.

IBM only recently made its commitment to openness. It has outlined parts of its network model, such as synchronous data-link control, since 1973. The company recently began emphasizing SNA's openness. IBM also has announced an Open Communications Architecture (OCA). It also plans to publish documentation about specific architectures and to hold classes to help other vendors design SNA products. This announcement covered 14 network architectures, including SDLC and emerging items such as LU 6.2.

One attempt to open SNA is the Systems Application Architecture (SAA). This lets vendors write a single software package that will run on IBM Personal Computers, Systems/3Xs, and mainframes. Vendors will not have to develop separate packages for each processor.

SAA might be a response to the success of Digital Equipment Corporation. DEC VAXes of all sizes run the same operating system and application software and are easily tied together. IBM products do not have this flexibility because the company supports a wide range of processors that cannot be easily connected.

In order to implement SAA, IBM had to select key network architectures. Some existing standards might never be included in SAA.

For example, IBM has decided that transmission-control protocol/internet protocol is needed only by a small group of users, such as universities and government. TCP/IP is not viewed as an appropriate networking solution for commercial users.

NETBIOS network interface software, which was introduced with the IBM PC Network, also might not be a part of SAA. NETBIOS is bound to the PC environment and cannot be used easily with other types of processors.

OSI is meant to be an interconnection point between networks, and in many cases this means proprietary networks provided by vendors such as IBM and others. IBM does not intend to abandon SNA and use the OSI protocols instead. OSI will be used as an interconnection mechanism to provide access outside of the SNA network.

IBM has stopped publishing interfaces in readily available manuals and now publishes them only in proprietary product manuals. It publishes and allows networking specifications and other vendors to develop SAA products. IBM does not license any code to these same vendors.

By not licensing code, IBM can control SNA and ensure a comfortable lead in developing new products. Because other vendors must rely on IBM for interfaces, IBM can control or change the interfaces as they wish.

Because there is a time lag between developing and publishing architectures, this provides a competitive edge. Many IBM architectures remain proprietary and closed. IBM has not opened up its wiring and cabling schemes, net management in the Token-Ring Network, or Rolm telephone systems sets. How open the PC Operating System/2 will be is another question.

Since about 1982 IBM's commitment to OSI has been more active. Participation in the standards process has grown, and IBM claims to be one of the few vendors other than DEC that has announced products that conform to all seven layers of the OSI model.

IBM's OSI products (TABLE 6-8) have been rated as equal and in some cases superior to DEC's compared to IBM's products. DEC's packet-switching products lack some features such as automatic restart and redialing capability and security.

IBM has sold only products that conform to the lower three layers of OSI in the U.S. Products that conform to the upper four levels have been available only to European customers.

OSI requirements have been greater in Europe, and in some cases these have been mandated by governments. The issue of incorporating SNA protocols in OSI has come up three times. When OSI was originally devised in the late '70s, IBM offered its SNA architecture as a blueprint for the OSI model, but this was rejected. IBM has twice offered its LU 6.2 protocols for inclusion in the OSI model, but this issue has not been resolved.

If LU 6.2 is included in OSI, third-party vendors would have to develop only one set on peer-to-peer protocols that would work with both SNA and OSI.

APPC/LU 6.2

Advanced Program-to-Program Communications (APPC), also called Logical Unit 6.2 or LU 6.2, is a communications architecture within IBM's Systems Network Architecture (SNA). It is particularly important for communications between microcomputers and other types of machines, such as mainframes and minicomputers.

APPC/PC is IBM's implementation of APPC for the personal computer. APPC/PC comes in two parts: the main body of APPC/PC and an application subsystem, which is a user-written extension to the main body of APPC/PC. Transaction programs (TPS) are required to make use of APPC/PC's services.

APPC/PC works with either a Token-Ring Network adapter card or an SDLC (synchronous data-link control) adapter card. With the Token-Ring card, IBM provides a small adapter handler program, which is APPC/PC's interface with the card that fits into the microcomputer.

Table 6-8. IBM OSI Products.

Layer	Product	Function
7	FTAM Programming Request for Price Quotation (PRPQ)	Supports MAP 2.1 FTAM on MVS in conjunction with MCS. Available in U.S.
6,7	X.400 Message-Transfer Facility	Message transfer agent between VTAM and X.400. Available in Europe in September 1988.
6,7	X.400 DISOSS connection	Conversion facility between DISOSS and X.400. Available in Europe in September 1988.
3,7	MAP communication Server (MCS)	Supports MAP 2.1 on Series/1 RPS. Available only in U.S.
3 to 7	MAP Application Server	Translates SNA protocols to MAP 2.1 protocol. Available only in U.S.
4,5,6	General Teleprocessing Monitor for OSI (GTMOSI)	Session manager that communicates with VTAM, FTAM, X.400, OTSS, OSNS, CICS, NCP, NPSI, TSO, MSNF, and IMS. Available only in Europe.
4,5	Open-Systems Transport and Session Support (OTSS)	MVS, VSE, or VM application that supports selected OSI transport and session-layer protocols. Available only in Europe.
3	Open-Systems Network Support (OSNS)	MVS or VM application that manages NPSI virtual circuits over packet-switched network. Available only in Europe.
1,2,3	Communications and Transmission Control Program (CTCP)	Application for initiating SNA sessions with NPSI gateway.
1,2,3	NCP Packet-Switching Interface (NPSI)	Front-end processor connection that provides SNA path information through packet-switched network for IBM mainframes.
1 to 3	Integrated X.25 adapter	Supports X.25 emulation on remote Series/1, System/88, and 5251 cluster controllers or local System/36 System/38.
1,2	MP-500 MAP interface	Lets IBM Personal Computers connect to IBM Token-Ring Networks running MAP 2.1. Available only in U.S.
1	Short Hold Mode/Multiple Port Sharing (SHM/MPS)	Hardware for the 4361 mainframe, System/36 and Series/1 to provide X.21 connection.

Key:
FTAM—File Transfer and Access Method
IMS—Information Management System
MAP—Manufacturing Automation Protocol
MSNF—Multisystem Networking Facility
NCP—Network-Control Program
RPS—Real-time Programming System
SNA—Systems Network Architecture

IBM's Network Basic Input/Output System (NETBIOS), which also might be present in the user's PC environment, can be thought of as functionally parallel to APPC/PC. NETBIOS has a separate service access point (SAP) (a logical software window) through which communications pass between it and the adapter card. There is no other contact between NETBIOS and APPC/PC at higher layers. Other software, such as IBM's PC LAN Program, can sit on top of NETBIOS (communicated through the NETBIOS interface) without interacting with APPC/PC.

Non-NETBIOS system software might use an approach that is similar to IBM. DOS is able to reach down into the local hardware and serve as a networking interface. DOS provides the networking interface to APPC/PC. APPC/PC, the PC LAN program, NetWare, NETBIOS, the adapter handler software, and DOS are all accessed through interrupts. By issuing the appropriate command, any application can gain the attention of the networking software at any time.

Because APPC is part of SNA, the SNA layers form the theoretical constructs for interface. This is shown in FIG. 6-6.

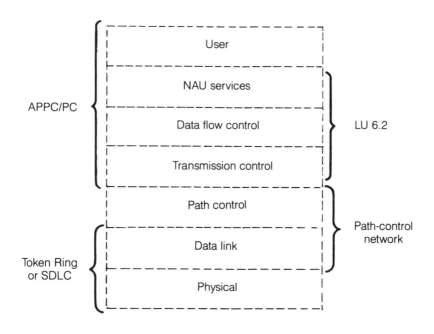

```
                    ┌─ ┌─────────────────────────────┐
                    │  │            User             │
                    │  ├─────────────────────────────┤ ┐
                    │  │         NAU services         │ │
          APPC/PC  ─┤  ├─────────────────────────────┤ │
                    │  │       Data flow control      │ ├ LU 6.2
                    │  ├─────────────────────────────┤ │
                    │  │     Transmission control     │ │
                    └─ ├─────────────────────────────┤ ┘
                       │         Path control         │ ┐
                       ├─────────────────────────────┤ ├ Path-control
                    ┌─ │          Data link          │ │   network
         Token Ring │  ├─────────────────────────────┤ ┘
         or SDLC   ─┤  │           Physical           │
                    └─ └─────────────────────────────┘
```

APPC —advanced program-to-program communications
LU   —logical unit
NAU   —network addressable unit
PC   —personal computer
SDLC —synchronous data-link control

Fig. 6-6. SNA Layers.

A Token-Ring of SDLC card provides physical and data-link functions. APPC/PC handles path control, transmission control, dataflow control, and NAU operations.

An APPC/PC software module has several ports, each offering access to a different network service. These access ports take the form of software "calls," or *verbs*. These verbs are the interface between the application software and APPC/PC. The set of verbs can also be referred to as the *protocol boundary* or the *applications-programming interface* (API).

APPC/PC verbs are divided into conversation verbs and control operator verbs (TABLE 6-9). There are three types of verbs within each division. The *conversation verbs* are used by transaction programs to communicate with other transaction programs. The *control operator verbs* are used by the application subsystem to control the local communications node.

Consider a possible conversation and assume that a transaction program (TP) is already running at each APPC/PC-equipped node.

A typical sequence of events follows. TP in node 1 issues an ALLOCATE verb. This causes APPC/PC to allocate a conversation locally and place an allocation request for the node 2 TP in the send buffer. TP 1 issues the SEND-

Table 6-9. APPC Verbs.

Mapped Conversation Verbs	Post_On_Receipt
MC_Allocate	Prepare_To_Receive
MC_Confirm	Receive_And_Wait
MC_Confirmed	Receive_Immediate
MC_Deallocate	Request_To_Send
MC_Flush	Send_Data
MC_Get_Attributes	Send_Error
MC_Post_On_Receipt	Test
MC_Prepare_To_Receive	
MC_Receive_And_Wait	**Control Operator Verbs**
MC_Receive_Immediate	Change Number of Sessions Verbs
MC_Request_To_Send	Change_Session_Limit
MC_Send_Data	Initialize_Session_Limit
MC_Send_Error	Reset_Session_Limit
MC_Test	Process_Session_Limit
	Session Control Verbs
Type-Independent Conversation Verbs	Activate_Session
Backout	Deactivate_Session
Get_Type	
Syncpt	**LU Definition Verbs**
Wait	Define_Local_LU
	Define_Remote_LU
Basic Conversation Verbs	Define_Mode
Allocate	Define_TP
Confirm	Display_Local_LU
Confirmed	Display_Remote_LU
Deallocate	Display_Mode
Flush	Display_TP
Get_Attributes	Delete

DATA verb. APPC/PC places the data in the send buffer. (If the send buffer were full, the data would be sent at this point.) TP no. 1 then issues a DEALLOCATE. This causes APPC/PC to discard its control information for the conversation and to place a deallocate indication in the send buffer. APPC/PC sends the contents of the send buffer and ends the conversation.

APPC/PC at the remote node (node 2) receives the allocation request from node 1 and strips it off, passing the data on. TP no. 2 receives the data by issuing a RECEIVE AND WAIT verb. It then receives the deallocate indicator by issuing another RECEIVE AND WAIT verb. TP 2 issues a DEALLOCATE, which causes APPC/PC at the node where TP 2 is running to discard its control information for this conversation.

Conversation verbs can be basic or mapped. *Basic conversations* use a header. The head field has 2-byte logical length (LL) portion, which specifies the total length of the logical record, including the header field. The header field can also contain an identification (I.D.) portion that specifies the type of data being sent. The LL portion and the I.D. portion form what is called the *LLID prefix*.

With a mapped conversation, the transaction program sends data with no header. APPC/PC adds the header automatically. APPC/PC is intended for distributed transaction procession, but there is no way for a PC programmer to define a transaction. This might be an attempt to maintain control at the mainframe. This limitation of APPC/PC functionally might be because APPC/PC is already a large program. It can require over 200 kbytes of main memory—almost a third of the memory available under MS-DOS versions 3.2 and earlier. Transaction-tracking methods typically require significant memory or disk space to store backup information, which is necessary to allow the node to undo aborted transactions.

The absence of a synchronization point* with APPC/PC can also be based on performance. Every time a program makes a change either in memory or on the disk, it must store all the existing information so that it can restore it if necessary. This data must be stored in memory or on the disk. Storing it on disk can double the number of disk write operations and affect performance.

In a microcomputer network, the network operating system can compensate for this function by providing a transaction-tracking function. LAN transaction tracking typically operates in a file server, which could have several megabytes of main memory and a large disk space.

APPC/PC allows one local program to emulate other local programs. Conversations that are requested remotely must be assigned to different emulated

---

*In APPC, a sync point allows two or more programs to establish points of consistence in their processing. These are called *sync* or *synchronous points*, to which they can all return if an error occurs in any program. This provides error recovery for distributed-transaction processing in which program is performing one part of a single logical unit of work. This function is not supported by APPC/PC.

programs. Conversations that are requested locally can all be assigned to one program or emulated program, or to multiple emulated programs with multi-tasking DOS (IBM's Operating System/2). Multiple remotely requested conversations can be assigned to multiple programs.

APPC/PC identifies programs with transaction-program identification numbers, of TPIDs. Locally requested conversations can be associated with one of several TPIDs. Remotely requested conversations must have different TPIDs. This is a formal distinction and does not affect the operation of the program running in the PC.

This is because the APPC architecture assumes a multitasking environment. APPC/PC was designed for single-tasking machines, in which one transaction program has to handle multiple conversations. APPC/PC solves this by allowing one program to pretend to be many programs.

Many APPC/PC control verbs are not defined by the APPC architecture. APPC/PC also has a passthrough feature that allows the application subsystem to define its own verbs. These verbs can be invoked using the same interrupts as for APPC/PC, so they act as an extension to the APPC/PC protocol boundary.

APPC/PC also offers an ASCII/EBCDIC conversion service that converts characters from the ASCII (American Standard Code for Information Interchange) character set normally used on IBM personal computers and the EBCDIC (Extended Binary Coded Decimal Interchange Code) character set typically used on larger IBM machines.

There is also a command that allows programs to enable and disable APPC/PC. Even when disabled, APPC/PC continues to queue incoming message units, but it only responds to verbs with an "APPCDISABLED" return code.

An important feature of APPC/PC is a hook into IBM's network management scheme. The TRANSFERDATA verb (which is not defined by the APPC architecture) allows an application to send network status and error information to a System Services Control Point, which typically is either a mainframe or communications controller. The error information is formatted as a network-management vector-transport (NMVT) request/response unit (RU). The ability to communicate with an SSCP using an NMVT RU allows APPC/PC to be compatible with other network-management software, such as NetView, which reports error and status information in much the same way.

IBM has been willing to share its SNA development work with OSI committees. The company has addressed network-management issues associated with multivendor environments. Network management is IBM's key to differentiating SNA from OSI.

The concepts of NetView and the OSI network-management standards are very close. The standards body could either create its own network-management tools or incorporate portions of NetView.

To attach its devices to an SNA network, competitors must layer IBM products on top of their own products. To support NetView, network-management vendors have to develop software to link their devices to NetView.

OSI CONNECTIVITY CASE STUDIES: DIGITAL EQUIPMENT CORPORATION

One of DEC's main strengths has been its product integration through DECnet. Almost everything the company makes can communicate over this network, and DEC has been providing connectivity to non-DEC systems. For example, the DECnet-DOS system allows DECnet interconnection with IBM Personal Computers through either asynchronous connections or direct connections to EtherNet. This implementation of DECnet allows an MS-DOS or PC-DOS system to function as a network node.

Another example of DEC's efforts to interconnect non-DEC machines is VAX/VMS services for MS-DOS. This facility allows users of MS-DOS-based personal computers with access to VAX/VMS files to print services network management and control services. VAX/VMS services are available to personal computer users with Microsoft Corp.'s MS-NET network, which uses the data-link layer protocols of Digital Network Architecture.

The Remote System Manager (RSM) allows centralized management of dissimilar VAX network nodes. RSM provides management for a collection of VAX/VMS computer systems, including single-user or multiuser MicroVAXes or VAXstations, on EtherNet. This also can include bridges to other EtherNet segments.

RSM also lays a foundation for extending such control to other DEC network segments, including non-VAX machines running operating systems other than VMS. For example, systems running Ultrix (DEC's version of Unix) have been interconnected through DECnet Ultrix. The development of enhanced services for Ultrix and other DEC environments that are functionally similar to VAX/VMS services for MS-DOS would allow interconnections to extend beyond DEC systems.

RSM allows increased efficiencies in system utilization, enhanced system and file security, and provides standardized backup and recovery routines.

VAXCLUSTERS

Several years ago, DEC announced the VAXcluster, which is a high-speed processor-to-processor bus that interconnects VAX computers and mass-storage units. It is based on DEC's Systems Communications Architecture, which is the internal architecture of the VAX/VMS operating system that supports clusters.

The VAXcluster supports a fully distributed file system. The original VAXcluster implementation enable VAX processors, including the 11/50, 11/780,

11/782, 11/785, and 8000 series to be connected with DEC's Hierarchical Storage Controller disk controller. This provided fault tolerance and access to shared data and system files. The architecture of the VAXcluster allowed every VAX processor in the VAXcluster to boot and run from the same operating system copy, which is kept in a central storage area.

In a VAXcluster, the combined processing power can be available to users as a service on their networks. When users connecting to the local network request a service, they are connected to a VAXcluster that supports this service. Through DEC's terminal servers, the user is transparently connected to any available processor on the service VAXcluster.

A heterogeneous cluster runs different versions of the VMS operating system and application software on each node. Each node is managed separately and shares access to centrally located data.

Distributed systems can also be heterogeneous, because they usually contain a mixture of different system software and hardware. They also comprise different application environments and can have different usage patterns for the same applications software.

Using VMS Version 4.6, with synchronization provided by Distributed Lock Manager VAXcluster system software, each of 100 jobs on 27 VAX processors can write to the same record in the same file simultaneously, without interfering with one another.

With RSM, DEC has been able to extend the VAXcluster concept to networks. VAXclusters were originally distance-limited because the 10 Mbps coaxial cable connections between cluster components had a maximum radius of 45 meters. DEC's Local Area VAXcluster extends the connections to include operation over EtherNet.

An extension of task-to-task interprocess communications mechanisms exist in the DECnet architecture. Much DEC software is now based on the concept of having servers that can reside on different network nodes. These nodes only serve requests from the network for access to their resources. In this context, the server is a multithreaded, multitasking program that can service many simultaneous network-access requests to one or many resources on that node.

Examples of such products include VAX Notes, the VAX/VMS multiuser teleconferencing system; the VAX VTX videotex system; and Datatreive, a data-query language and report writer.

DEC has made it possible for tabletop VAX workstations to be configured in a wide variety of ways. For example, a VAX2000 workstation can be run in a diskless fashion in which the operating system is downloaded from the network.

Because the swapping and paging activity for the operation of a VAX/VMS system can be costly when down over the network, a workstation can be configured with a local disk that can be used for swapping and paging. Another possible configuration is for the operating system and the application software to reside on a workstation's local disk, with the local-area VAXcluster providing an access path to centrally stored data.

Local-area VAXclusters that are managed in a homogeneous manner are especially well-suited to work groups in which ease of access to central data and ease of system management are important.

Single-point management of distributed systems for up to several hundred homogeneous, geographically dispersed nodes is also possible. This could include such capabilities as downloading and management of operating systems, installment list verification, performance out-testing, network bridge management, fault management, loop-back and remote diagnostics, error logging, accounting, security through encryption and key distribution, and test/certification/release for application software.

DEC is maintaining its lead in networking by continuing to introduce integrated tools for distributed-system management. DEC's networking strategy provides users with an architecture that will continue to offer increased connectivity. As a result, if users require network solutions that DEC does not have now, it is likely that they could be added in the future.

For organizations that must manage collections of MS-DOS, CP/M, Unix, and VAX/VMS systems, the time might be near when it will be possible to command printing on the printers that are closest physically on the network. Advances in DEC's networking technology also are likely to include generic file servers and generic disk spare servers.

NETWORK MANAGEMENT

Using the DECnet Monitor, a DECnet user can establish a central control point for network management. The DECnet Monitor maintains a data base of network information that is presented to a user through displays and interactive graphics. For example, if a point-to-point link should fail, that failure will be recorded at each end of the circuit. If there is more than one failure, the DECnet Monitor will automatically sort out duplicates ignoring any redundant information.

A network-control program (NCP) keeps counters on traffic volume and error statistics. It also computes message and error rates based on these statistics. The DECnet Monitor also detects increasing error rates and provides displays which highlight such trends.

The DECnet Monitor system creates a model of the network using a data base of the relationships of the network components. Special monitor software does not have to reside in each node. The monitor can collect information about any node in the network. This differs from the IBM's approach. IBM's Response Time Monitor software is required in remote devices that supply statistics to NetView.

DEC does not yet have an overall umbrella product like NetView to integrate network-management products. Instead, it supports a series of network-management products.

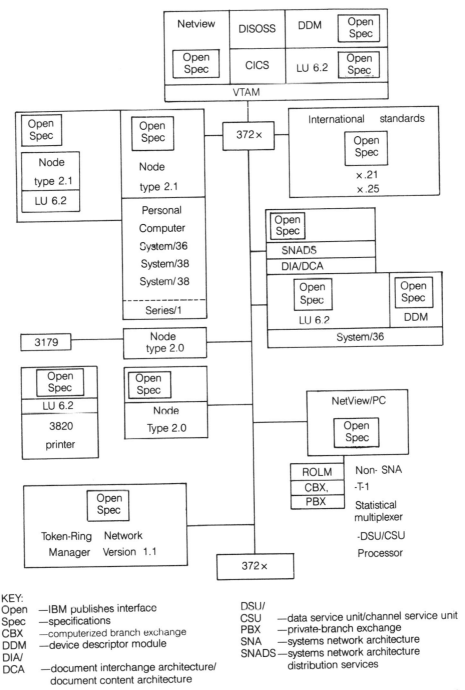

Fig. 6-7. IBM Open Specifications for communications and network management.

KEY:
Open —IBM publishes interface
Spec —specifications
CBX —computerized branch exchange
DDM —device descriptor module
DIA/
DCA —document interchange architecture/
document content architecture

DSU/
CSU —data service unit/channel service unit
PBX —private-branch exchange
SNA —systems network architecture
SNADS —systems network architecture
distribution services

Cable/Facilities Management (C/FM) is VAX/VMS software that manages voice and data cabling, PBX capacity planning and work-order functions. C/FM includes cable management, station software features, PBX common equipment, inventory, and work-order entry modules.

PBX/Facilities Management (P/FM) tracks line costs and bills from call recording for PBXs. It can produce invoices and equipment chargebacks.

VAX Ethernim is an EtherNet maintenance program that runs under DECnet software on a system configured as DECnet node.

Ethernim builds a database of information about each node directly connected to the EtherNet cable. It also can test EtherNet connections to all nodes on the network except MicroVAX I and VAXstation I to Digital-Network-Architecture network applications and user layers for any node running DECnet Phase IV on the EtherNet.

Ethernim runs as a layered product on the VAX/VMS operating system. It can be operated from local terminals using Regis graphics for the network under test.

The Remote-System Manager (RSM) is VMS-layered software that allows a network manager to perform software distribution, installation, backup, and update tasks for DECnet EtherNet nodes.

The server provides services that have been defined in the server's database and are connected to the same EtherNet. Remote Bridge-Management Software (RBMS) is used with the LAN Bridge 100. RBMS can disable selected bridges or block traffic to isolate a specific segment of an extended network.

DEC uses a subset of the IEEE 802.1 Management Protocol to communicate between the VAX host and the targeted bridge, but RBMS provides more functions and control than the LAN Bridge 100. The LAN Traffic Monitor (LTM) captures information on traffic from a bridged network. A LAN Bridge 100 (which is downloaded with the monitoring software) is attached to the EtherNet cable and transmits information to the LTM host software program, which can be located on any VAX in the extended local network.

The LTM collects statistics on EtherNet traffic regardless of the higher level protocol used. Customers using DECnet, Transmission Control Protocol/Internet Protocol, or Xerox Network System can analyze the higher level network-protocol performance. LTM also provides real-time and historical network performance graphs for EtherNet segments.

OSI CONNECTIVITY CASE STUDIES: HEWLETT-PACKARD

Hewlett-Packard (HP) has been involved in computer networking since the early 1970s and has made a major effort to develop and implement standards and open systems. Because networking is a key technology in computing, any company that wishes to play a role in the marketplace must make the investment. Networking is now one of the two top programs for HP. Multivendor connectivity is the driving force behind their AdvanceNet networking strategy, which is really a networking and systems-integration strategy and not an architecture like SNA or DECnet.

With its traditions in manufacturing and engineering environments, HP has been heavily involved in Manufacturing Automation Protocol and Technical and Office Protocol standards activities. The company sells products based on MAP 2.1 specification.

In its factory products, HP has been using broadband technology developed by Ungerman-Bass, Inc. and Industrial Networking, Inc., which is a joint venture between Ungermann-Bass and General Electric Company. HP TOP 3.0 products have been developed by its Information Networks Group. In Europe, HP has announced an X-400 messaging product for the Unix environment.

OSI IMPACT

In order to deliver MAP and X.400 products, HP has had to develop full seven-layer protocol stacks based on the ISO Open Systems Interconnect (OSI) model. HP is using TCP/IP as its transport-level protocol. On top of TCP/IP, HP offers its own network services as well as the applications traditionally associated with TCP/IP: Telnet for remote logon, File-Transfer Protocol, and Simple Mail-Transfer Protocol.

TCP/IP is a network protocol developed by the Department of Defense. It can provide companies with a common system for interconnecting different brands of computers to transfer files and mail messages.

The protocol can be used in layers 3 and 4 of the International Standards Organization's Open Systems Interconnect network model. It provides the error detection and correction capabilities required for accurate data transmission from one device to another. The file transfer and other capabilities are achieved using the applications that have been built on top of TCP/IP. These include the File-Transfer Protocol, TELNET, which allows for remote log onto different systems, and the Simple Mail-Transfer Protocol, which provides electronic-mail services.

HP also has a version of Sun Microsystems, Inc.'s Network File System (NFS). NFS provides file sharing across systems connected on a local network, and it can be used on top of TCP/IP. This development was for the HP 9000 line.

HP has worked with the Wollongong Group of Palo Alto to provide TCP/IP applications for its minicomputer line. A version of the TCP/IP application software is for the HP 3000 line. To guarantee interoperability, HP has tested its implementation of TCP/IP against several other vendors' implementations. By swapping out lower layer protocols while maintaining its application-level networking services, HP was able to move to an industry-standard protocol stack in such a way that customers did not have to change their code.

Although HP supports EtherNet on its 9000 series, it has not supported it on the 3000 line. HP has opted to support IEEE 802.3 for local networking instead of EtherNet. While IEEE 802.3 and EtherNet are close, they are not close enough. They can share the same medium, but they cannot talk to each other.

HP has a joint project with M/A-Com Telecommunications Division, a subsidiary of M/A-Com, to sell and support private X.25 networks based on M/A-Com's line of packet-switching products. HP uses 802.3 hardware developed by 3Com Corp.

HP is an active participant in a wide range of standards bodies such as the IEEE 802.3 work group and is vice-chairman of the Accredited Standards Committee (ASC) X3. About one-third of the approximately 75 X3 subcommittees and task groups are working on data communications and networking areas, including development of OSI-compatible standards. ASC standards work is done in accordance with guidelines set by the American National Standards Institute. HP was also instrumental in forming the Corporation for Open Systems and is a member of its executive committee.

HP is also involved in the National Bureau of Standards special-interest group on upper layer network architecture and of the Technical Committee 32 task group of the European Computer Manufacturers Association, which addresses a range of networking and system-integration issues. HP is also active in the Consultative Committee on International Telephone and Telegraphy, as well as in European and Asian groups such as the Standards Promotion and Applications Group, the Interoperability Technology Association for Information Processing, the Promoting Conference for OSI, and the European MAP Users Group.

HP also participates in the X.12 committee, developing electronic-document-interchange (EDI) standards. HP offers a range of direct connect and remote Systems Network Architecture and Binary Synchronous Communications connections, providing terminal emulation and file transfer.

HP offers links to IBM's DISOSS and Professional Office Systems and supports IBM's Document Content Architecture and Document Interchange Architecture. The company introduced its first LU 6.2 product in conjunction with its DISOSS link, which uses the LU 6.2 product to make the connection between the HP 3000 and the DISOSS host.

HP SNA products act as if they were nodes on an SNA network. HP also provides links from personal computers to its HP 3000 and to IBM hosts.

A product called HP Information Access allows personal computer users to pull data from both IBM and HP hosts in what has been described as a seamless, transparent manner.

HP also has worked with Cullinet Software, Inc. to develop the capability to access database-management systems on IBM mainframes.

While DEC has emphasized the use of gateways between its environment and IBM's, HP has taken a plug-compatible stance. In connecting to DEC, HP offers TCP/IP or a product called Network Services for the DEC VAX Computer. This product implements HP's Network Services on the DEC VAX/VMS family of computers and permits file transfer between the DEC systems and HP minicomputers. Network Services essentially makes a DEC machine look like an HP machine.

HP does not provide many special products for linking into any other vendors's environment. It relies on SNA and BSC connections to communicate with systems from vendors such as Unisys Corp. and TCP/IP for links to companies such as Sun Microsystems and Apollo Computer.

Adhering to standards and connecting into other environments are not only factors that define the openness of a vendor's environment. Third-party hardware and software vendors also must be able to connect for true openness to exist. To aid third-party vendors, HP offers three levels of application interfaces in its networking architecture.

A link-level interface that resides on top of TCP/IP provides program-to-program communications across systems. A services-level interface provides developers with file-transfer protocols, remote database access, and other capabilities needed to construct network applications.

A third interface provides distributed application services to users and developers can create distributed applications. HP's NetDelivery provides asynchronous store-and-forward functions across the network.

HP also provides specifications for its network-management system to allow modem and terminal suppliers and other vendors to participate in HP's management scheme.

TABLES 6-10 and 6-11 summarize the OSI activities and connectivity support.

OSI CONNECTIVITY CASE STUDIES: DATA GENERAL

Data General (DG) plans full Open Systems Interconnect (OSI) compliance and an ongoing commitment to support de facto standards such as IBM's Systems Network Architecture (SNA) (see TABLE 6-12). It has also introduced PC*I, a network plan integrating IBM Personal Computers and compatibles into DG's product family.

DG's PC*I product provides an example of the twin strategy of implementing SNA while tying in OSI. With PC*I's networking architecture, users have a choice of running 802.3 EtherNet, ThinWire EtherNet, or Starlan at layer 1 of the OSI model. At layer 2, DG supports 802.2. TABLE 6-13 shows DG's other network products.

DG has implemented the International Standards Organization's (ISO) Internet Protocol (IP) at layer 3 and transport protocol (TP) at layer 4 via its Workstation Transport System (WTS) on the terminal side and its Xodiac Transport System (XTS) on the host side. Network Basic I/O System provides

Table 6-10. Open-Systems Interconnect (OSI)
and Other International Standards Supported by Hewlett-Packard.

	Present					Proposed			
Services OSI layers 6,7	TCP/IP applications	HP network services*	MAP 2.1	X.400	HP network services*	MAP 3.0	TOP 3.0	X.400
	Berkeley services		Genetic OSI services		Generic OSI services			
Transport OSI layers 4,5	TCP/IP		OSI transport		OSI transport			
Links OSI layers 1,2,3	802.3 802.4, X.25, Point-to-point				802.3 802.4, X.25, Point-to-point ISDN			

*Proprietary

Table 6-11. Multivendor Connectivity Provided by Hewlett-Packard.

Present Connectivity for IBM

- HPOffice Connect to DISOSS—link from HPDesk Manager to IBM's Distributed Office Support System for HP 3000
- HPOffice Connect to PROFs—link from HPDesk Manager to IBM's Professional Office Systems for HP 3000
- SNA Server and SNA Server Access—allows an HP 3000 to act as a Systems Network Architecture gateway to multiple HP 3000 Systems
- HP LU 6.2 Base—provides connection between HP 3000 and IBM DISOSS host
- SNA Interactive Mainframe Facility—emulates main features of IBM 3270 control unit using PU 2 and LU 1, 2, and 3 protocols; provides pass-through logon to IBM mainframes as well as a programmatic interface for program-to-program communications; binary synchronous version also available for HP 3000
- Programmatic Mainframe Facility—for program-to-program communications for HP 1000
- SNA Network Remote Job Entry—batch-oriented emulation for HP 3000, 1000, and 9000
- Remote Job Entry and Multileaving Remote Job Entry—binary synchronous/batch-oriented links used in conjunction with BSC Link HP 3000, 1000, and 9000
- SNALink—manages physical link to host, implements protocols in SNA's lower three layers, and provides PU 2-type connection for HP 3000
- SNALink/3270—provides 3270 emulation and file transfer for remote personal computers
- HP IRMA—provides personal computers with 3278 emulation and file transfer via direct-connect access to IBM hosts
- HP-UX SNA 3270—interactive emulation product for HP 9000 that can be used either as a gateway or by a stand-alone system.

Table 6-11. (Continued)

Proposed Connectivity for IBM

- LU 6.2 Application Interface—for the development of HP 3000 applications capable of program-to-program communications with IBM applications.
- Reverse pass-through capability—will allow users on IBM terminals to access HP minicomputers
- SNA-to-X.25 conversion software
- Testing of HP StarLan over IBM type-3, unshielded, twisted-pair wiring.
- Testing of third-party Token-Ring Network Interface cards for use with the HP Vectra PC
- An 802.3-to-802.5/StarLan-to-Token-Ring bridge
- Support for IBM's Network-Management Architecture
- SNA 3770 batch emulation for HP 9000 family; will be configured either stand-alone or as a gateway
- Support for IBM PU 2.1 and Systems-Network Architecture Distribution Services for HP 3000

Present

- Digital Equipment Corp
- Network Services for DEC VAX Computer—provides network file transfer between VAX/VMS and HP 1000, HP 3000, HP 9000 Series 300, Series 500, and Model 840 computers.
- Transmission Control Protocol/Internet Protocol Applications
- Unix-based computers, including AT&T 3B line
- Berkeley TCP/IP services
- Honeywell, Inc.
- RJE emulation and file transfer
- HP OSI/Session-Transport—runs on top of X.25 and provides programmatic and interactive session-level access
- Manufacturing Automation Protocol File Transfer and Access Method (FTAM) connection
- Unisys
- RJE and Multileaving RJE emulation and file transfer
- Burroughs, Control Data, Data General, NCR, Siemens AG, Tandem RJE emulation and file transfer
- Cray Research
- Hyperchannel Data-Link Interface—provides direct connection between HP and Cray
- Apollo Computer
- MAP FTAM connection
- Sun Microsystems
- TCP/IP applications
- Wang Laboratories
- RJE emulation and file transfer
- Document-conversion utility

Table 6-12. DG's Migration to ISO.

Present

OSI reference model	Xodiac minicomputer architecture	PC integration architecture
Application	Xodiac network servers FTA MTA VTA RMS	PC Network servers MVNET WVTA
Presentation	RDA RIA	NETBIOS
Session		
Transport		ISO TP class 4
Network	Xodiac Transport System (X.25)	ISO IP
Data link	802 local area	
Physical	Networks HDLC, SDLC	Links

Proposed

Xodiac non-ISO network servers	PC Network servers	OSI applications	Xodiac ISO network servers FTAM, X.400, VT	Net management
	NETBIOS	ACSE, ROSE, RDA, TYSE, session, presentation		
Xodiac Transport Protocol (X.25)		ISO TP Class 4 ISO TP Class 0 (for X.400)		
		ISO IP		
	Links			

Key:

ACSE—associated control service element
FTA—File-Transfer Agent
FTAM—File Transfer and Access Method
HDLC—High-level data-link control
ISO—International Standards Organization
MTA—Message Transfer Agent
MVNET—MV Server for MSNET
NETBIOS—Network Basic I/O System

RDA—Remote Database Access
RIA—Remote Infos* Agent
RMA—Resource-Management Agent
ROSE—Remote-operations service element
TPSE—transaction processing service element
VTA—Virtual Terminal Agent
WVTA—Workstation Virtual-Terminal Agent
*Infos—DG proprietary remote database-management system

Table 6-13. DG's Internetwork Products.

Present Products

Full internetwork system based on X.25, supporting:

- File transfer
- Message handling
- Virtual terminal
- Transparent system resources, access and management
- Remote database access

Operates over IEEE local-area networks, wide-area synchronous links and through Public Data Networks.

Proposed Products

Full internetwork system based on ISO protocols, supporting:

- File transfer
- Message handling
- Virtual terminal
- Remote data access
- Remote procedure call

Operates over IEEE local-area networks, wide-area synchronous links and through public data networks. Uses Xodiac services and X.25 to communicate with previous revisions of Xodiac systems. Tuser, Data General's implementation of the Transport Service Interface, provides selection of, and access to, multiple transport protocols.

---

the standard application interface within the network architecture, allowing users to run applications written for IBM's PC Network.

DG's XTS/SNA backbone software allows DG systems running Xodiac networking software to communicate over an IBM SNA network.

Backbones are an approach to structured networking. In this concept equipment can be grouped functionally or grouped to provide another hierarchical layer. In a backbone network, each node, rather than being a single computer or device, can be the entry point into another network. This arrangement allows workstations allocated to specific projects to operate on their private network in addition to having access to an overall network. This overall network (backbone) can be the same type of network used to interconnect the workstations.

A backbone is a useful layer in a hierarchy of networking. The local network continues to be under the control of the local users and is tailored to meet their needs. The backbone provides the link to other areas and requires no special knowledge of those areas by the local users.

In addition to supporting Document Interchange Architecture, Document Content Architecture, DISOSS, Professional Office System, and SNA Directory Services via the Comprehensive Electronic Office automation product, DG's XTS/SNA lets DG systems running Xodiac-networking software communicate with one another in an IBM SNA environment. DG also supports PU 2.1

and LU 6.2, which allow for peer-to-peer communications. Using DG's LUG.2 product, users can establish sessions between a subsidiary DG station and an IBM mainframe. Generally, this is done in a DISOSS environment. DG offers IBM emulation, including 3270, 2780, 3780, HASP, 3770 Remote Job Entry, and 3289 printer emulation.

Within the OSI model, DG currently supports 802.3 and 802.4 standards at layers 1 and 2. Higher up, DG supports X.25, X.29, and X.400. The company plans to support 802.5 token-ring standards, OSI/File Transfer and Access Methods, Technical and Office Protocol, and Manufacturing Automation Protocol 3.0.

DG's OSI strategy is to implement the mode in ways that involve layering ISO protocols at the transport layer and then using de facto standards to operate at the upper layers. DG will be supporting Electronic Data Interchange and is supporting the Department of Defense's Transmission-Control Protocol/IP to provide functionality for layers 4 and up.

Users can run traditional TCP/IP applications such as Telnet, File-Transfer Protocol and Simple Message-Transfer Protocol using Xodiac application services. These services are what DG calls *agents* and include DG's Resource Management Agent, Virtual Terminal Agent, File-Transfer Agent and Remote Database-Management Agent.

DG's plans to migrate its Xodiac architecture to OSI compliance call for an enhancement to its X.25 product, which will migrate from 1980 specifications to 1984 specifications. This will allow DG to use Network Service Address Points for addressing. DG will also replace X.25 on local-area networks with TP/IP, the ISO equivalent of TCP/IP.

Other planned OSI connectivity projects include support for OSI Session, OSI Presentation, and OSI Associated Control Service Elements. This includes FTAM, Virtual Terminal Survey, Remote Operations Service Element, and Transaction Processing Service Element as well as Integrated Services Digital Network. DG also plans to use MSNET as its basis for connectivity between personal computers and DG superminicomputers. For users connecting to DEC systems, DG offers TCP/IP as well as a product called DEC systems, DG offers TCP/IP as well as a product called CEO Document Exchange III that permits file swapping. DG also provides connection to Wang through CEO Document Exchange I, which allows DG systems to exchange documents with Wang word-processing systems and lets users mail documents via Wang's Mailway.

DG'S Blast file-transfer software provides connections to systems from HP, Tandem, Apple, and IBM; TCP/IP offers links between DG machines and those from Unisys, Sun Microsystems, Apollo, HP, Prime, and Tandem.

Connectivity is also provided by third-party vendors. The Cullinet Software Information Database lets CEO users access information in a Cullinet database via an SNA link. It also links DG systems to Cullinet Information Databases run-

Table 6-14. Data General Connectivity.

IBM

- DG/PC*1—connects IBM Personal Computers and Eclipse/MV servers using IEEE 802.3 thick, thin, or StarLan Configurations
- CEO Connection—supports asynchronous file transfer between an IBM Personal Computer and an Eclipse/MV host running CEO software
- DG/SNA—allows Eclipse computers to appear as an SNA PU 2
- DG/SDLC—provides capabilities for access to IBM systems using SNA.
- XTS/SNA (backbone software)—lets DG systems running Xodiac networking software communicate over an IBM SNA network
- X.25 Data-Link Control Software—allows users to access SNA networks via X.25
- DG/Blast
- CEO Document Exchange Architecture:

    1. allows a DG system with CEO to emulate an IBM office system node
    2. lets CEO users exchange documents and mail with DISOSS users
    3. lets CEO users store and retrieve documents in the DISOSS central library
    4. Converts documents on the DG system without demand on the IBM host
    5. users IBM's SNADS, DIA/DCA, PU 2.1, and LU 6.2 distributed-network architectures

- PROFS—supports exchange of electronic mail and documents between CEO and PROFS using DISOSS Exchange Architecture
- RUE80—IBM 2780 or 3780 Remote-Job-Entry station emulator
- RCX70 (IBM 3271 Remote Cluster Executive)—lets Eclipse systems emulate an IBM 3271 cluster controller
- HASP II—IBM HASP II RJE workstation emulator
- SNA/3270—emulates the IBM 3270 Information Display System in a DG DG/SNA environment and lets DG users access an IBM host directly from their DG workstation. SNA/3270 lets an appropriate DG terminal appear as an IBM 3278 terminal to the SNA network.

Future

- IBM Token-Ring local-area network

Digital Equipment Corporation

- CEO Document Exchange III
- DG/X.400
- TCP/IP

Hewlett-Packard

- DG/X.400
- TCP/IP
- DG/Blast
- IEEE-488-bus—factory and scientific networking bus standard

Prime

- TCP/IP

Sun Microsystems

- TCP/IP
- DG/UX—includes an implementation of Sun's Network File Systems

Tandem

- CEO Document Exchange
- DG/Blast
- DG/Gate—terminal-emulation software
- TCP/IP

Unisys

- TCP/IP
- BSC

Table 6-14. (Continued)

- SNA/3270 emulator

Wang Laboratories
- CEO Document Exchange I—lets DG systems with CEO exchange documents with Wang word-processing systems

Apollo
- TCP/IP

Apple
- DG/Blast—enables asynchronous file transfer over dial-up or dedicated telephone lines

Cray Research
- Hyperchannel

Key:
BSC—Binary Synchronous Communications
CEO—Comprehensive Electronic Office
DCA—Document Content Architecture
DIA—Document Interchange Architecture
PROFS—Professional Office System
SDLC—Synchronous data-link control
SNA—Systems Network Architecture
TCP/IP—Transmission Control Protocol/Internet Protocol
XTS—Xodiac Transport Services

---

ning on an IBM mainframe. KAZ Business Systems provides a Macintosh interface to minicomputers and mainframes.

TABLE 6-14 summarizes connectivity with other vendors' equipment.

OSI CONNECTIVITY CASE STUDIES: APPLE

AppleTalk is part of Apple's overall communications strategy, called the Apple Communications Framework (ACF). ACF is the framework that defines workstation technology, network strategy, printing strategy, internetworking, and multivendor integration strategy.

AppleTalk is a full, seven-layer protocol that can be used to link the Macintosh to DEC and MS-DOS environments. Third parties also provide gateway devices that link AppleTalk to SNA and provide DECnet and TCP/IP support.

Apple's policy is to build enabling technologies and work with third parties to make services emerge. For a list of third-party products, see TABLE 6-15.

Apple has many products planned. Current multivendor communications capabilities include the AppleTalk personal computer card, which lets IBM Personal Computer users connect to AppleTalk networks. They are limited to printing documents on the LaserWriter. Think Technologies InBox/PC provides transparent Macintosh-to-IBM Personal Computer electronic mail and file transfer. IBM Personal Computer users who want to access the AppleShare file server can use AppleShare PC.

Tangent Technologies' Share also provides this capability. Several third-party options exist to connect Macintoshes via EtherNet, and Apple has its own 802.3-compatible EtherNet adapter. This was developed with 3Com Corporation. The EtherTalk card for the Macintosh II provides AppleTalk support on an EtherNet at 10 Mbps and will work with both the Macintosh operating system and Apple's AU/X Unix operating system. Apple also plans IBM Token-Ring Network support.

Apple's multitasking operating system, Multifinder, runs on most Macintosh models and allows multiple programs to run simultaneously, including communications applications. The operating system allows a user to work on one application in the foreground, for example, while in the background the Macintosh could be exchanging files with another machine.

The Multifinder multitasking operating system lets users switch easily among different applications, share data among applications, and run lengthy operations, such as hard disk backup, in the background. The program also makes it possible to pull information from MS-DOS based applications.

Multifinder's major impact might lie in the type of applications developed for it. Developers could build multipurpose applications that run communications processes in the background. Multitasking could result in applications that bring the mainframe world together with the desktop computer and make it appear seamless to the end user.

Multitasking also strengthens communications because one workstation can serve as a supervisor that allocates tasks. If a task is better performed on a workstation somewhere else in the network, that workstation can delegate the task to that piece of hardware.

Multifinder runs on all Macintosh II, Macintosh SE, and Macintosh Plus personal computers.

Apple products also include the Appleline 3270 Protocol Converter, which lets Macintoshes share information with IBM mainframes via terminal-emulation but lacked file-transfer capability. Now AppleLine 3270 File-Transfer software can transfer files between Macintoshes and the 3270 environment via the Appleline Protocol Converter.

AppleTalk is a typical 7-layer, ISO-oriented network architecture. The addition of AppleTalk Session Protocol and Appletalk Filing Protocol (AFP) adds utility with standard protocols for file service and multiuser applications.

AppleTalk is media-independent. Upper layer protocol stacks can be implemented on top of a range of physical and link-layer implementations other than the AppleTalk Personal Network, including EtherNet or token-ring networks.

Several protocol stacks can talk to lower layer implementations concurrently. Among the available cabling options are a DuPont fiberoptic scheme that supports AppleTalk at 230.4 kbps, and Faralon Computing's PhoneNet, which lets users run AppleTalk over nonshielded twisted pair. DuPont also is working on higher speed fiber-based configurations for bridging smaller, slower networks.

The AppleShare file-server software, AFP, lets AppleTalk-linked Macintoshes share a central hard disk and provides a variety of file and directory security options that improve on those of third-party servers such as TOPS. CENTRAM'S TOPS program allows transparent file sharing between Macintoshes, IBM Personal Computers, and Unix-based computers.

Apple also has purchased a minority interest in Touch Communications, a maker of versions of OSI software for IBM Personal Computers and VAX-VMS computers. Touch's OSI products comply with the Manufacturing Automation Protocol/Technical and Office Protocol draft Version 3.0 specifications. Touch will work with Apple to develop OSI software that will take advantage of the Macintosh's icon and menu-driven commands to access and transfer files from a Personal Computer or VAX over a TOP network.

Apple will also support TCP/IP using the AU/X operating system over the EtherTalk network. Under the Macintosh operating system, third-party products provide TCP/IP support, including access to Telnet and File-Transfer Protocol services.

Apple is also supporting ISDN philosophically. From the Mac frame of reference, ISDN is just a fast serial port. Apple also announced a facsimile modem called the AppleFax Modem. The AppleFax Modem is a 9.6 kbps asynchronous modem for sending and receiving files with standard facsimile machines or with other Macintoshes.

Available software gives IBM Personal Computers access to an AppleShare file server. File-translation software lets Macintosh users work with MS-DOS.

Apple File Exchange software handles the document translation between MS-DOS and Macintosh programs. DOS files on a 5.25 inch floppy disk fit into an Apple 5.25 inch disk drive for translation by the file-exchange software. The software is bundled with Macintoshes.

Apple also has announced a software engine called HyperCard that simplifies program development. Applications developed under HyperCard can be customized to accommodate individual working methods and special-purpose tasks such as teaching programs.

TABLE 6-15 shows the connectivity provided for Apple products. The products listed are representative of connectivity options available from other vendors.

OSI CONNECTIVITY CASE STUDIES: WANG LABORATORIES

Wang Laboratories is committed to providing communications products that are compatible with international Open Systems Interconnect standards. Standards are not an end in themselves at Wang; they are viewed as the means to providing information solutions across vendor lines.

Wang's approach toward standards is one of protecting customer's investments in hardware, networks, and software. Wang's view is that OSI implementations do not offer the features and functions found in proprietary commu-

Table 6-15. Apple Connectivity Provided by Apple.

I. PRESENT CONNECTIVITY
- **IBM Apple PC 5.25 disk drive**—MS-DOS drive for Macintosh SE and II
- **Apple File-Exchange Software**—document conversion between leading Macintosh and MS-DOS word processors. Included with Apple 5.25 disk drive
- **AppleTalk PC card**—provides IBM Personal Computer access to AppleTalk printing, mail, and file services
- **AppleLine Protocol Converter**—converts from synchronous to Systems Network Architecture or asynchronous. Provides IBM 3278 Model-2 terminal emulation. Copy and Paste, Copy Table supported by MacTerminal software
- **AppleLine 3270 File Transfer**—software application for AppleLine provides file transfer from TSO and CMS environments. Uses Irma host file-transfer software
- Digital Equipment Corp.
- **AppleTalk for VMS**—implementation of AppleTalk network protocols under VAX/VMS. Development environment for AppleTalk Services in VMS
- **MacTerminal**—VT100 terminal-emulation software

II. PROPOSED CONNECTIVITY
- IBM
- **AppleShare PC**—allows file sharing between Macintosh and MS-DOS computers
- **AppleShare for MS-DOS**—AppleShare file-server client software for MS-DOS personal computers provides transparent information sharing with Macintosh
- Support for LU 6.2, PU 2.1, 3270 Enhanced Connectivity Facility, and Token-Ring Network
- Digital Equipment Corp.
- **EtherTalk expansion card for Macintosh II**—EtherNet support for Macintosh II

III. CONNECTIVITY PROVIDED BY OTHER VENDORS
PRESENT CONNECTIVITY
- IBM
- Avatar Technologies
- **MacMainframe DX**—Systems Network Architecture/asynchronous protocol converter for Macintosh 512 and Mac Plus. Supports terminal emulation and TSO, CMS, CICS file transfer
- **MacMainframe SE**—3278 expansion card for Macintosh SE. Emulates 3278 Model 2 and 5, TSO, CMS, CICS file transfer using Avatar host software
- Centram Systems
- **Transcendental Operating System**—AppleTalk network file server for MS-DOS and Macintosh computers
- DataViz
- **MacLink Plus**—asynchronous file transfer between popular MS-DOS personal computers and Macintosh applications
- KMW Systems
- Series II and III Twinax Protocol Converters
- **S/3XLink Macintosh application software**—5251, 5191 terminal emulation and file transfer
- Lutzky-Baird Associates
- **UltraOffice**—networks IBM Personal Computers and Macintoshes via Unix
- Tangent Technologies
- **Tangent Share**—file server for Macintosh and MS-DOS computers
- Think Technologies
- **InBox/PC and InBox/Mac**—electronic mail and file transfer between Macintosh and MS-DOS personal computers on AppleTalk, Token-Ring
- 3Com
- **EtherSeries Enhanced Macintosh, 3Plus Macintosh**—provides parity network services to Macintosh and IBM Personal Computer
- TriData

Table 6-15. (Continued)

- **Netway 1000A**—AppleTalk-to-SNA gateway
- Wall Data
- DCF II protocol converter
- **MacBlue 5251 and FTX host software**—5251, 5291 terminal emulation, file transfer, and Synchronous Data Link Control connect
- Digital Equipment Corp.
- **Alsa Systems**

    1. **AlsaTalk**—AppleTalk network file, terminal, and print service
    2. Dove Computer
    3. **FastNet**—Macintosh small computer system interface (SCSI) to EtherNet Intelligent controller

- **Kinetics**

    1. **Fastpath**—AppleTalk personal network to EtherNet cabling programmable gateway. Supports AppleTalk, Transmission Control Protocol/Internet Protocol. DECnet support planned
    2. **EtherSC, EtherSE**—direct SCSI and Macintosh SE connection to EtherNet. Supports AppleTalk, TCP/IP, DECnet support planned
    3. **AppleTalk Q-BUS expansion card**—direct AppleTalk personal network connection for Micro VAX II

- **Odesta**

    1. **Helix VMC**—Macintosh multiuser database can run on Macintosh or VAX (under AppleTalk for VMS)

- Pacer Software
- **PC Link**—Macintosh/VAX network file, terminal, and print service
- Peripherals Computers & Supplies
- **Versatem Pro**—VT100, Tectronix graph terminal-emulation software
- 3Com
- Macintosh II EtherNet expansion card
- Direct Macintosh II EtherNet expansion card
- **White Pine Software**

    1. **Mac 240**— VT100, VT220, VT 240 terminal-emulation software
    2. **Regie**—Macintosh-to-DEC 240 Regis graphics translator

- Hewlett-Packard Co.
- **Tymlabs**

    1. **MAC2624**—HP 3000 and 1000 terminal-emulation and file-transfer software

- **Walker, Richer & Quinn**

    1. **Reflections for the Macintosh**—HP 3000 and 9000 terminal-emulation and file-transfer software

- Data General Corp.
- KAZ Business Systems

    1. **MacDasher**—file-transfer and DG D210 terminal-emulation software

- Prime Computer, Inc.

Table 6-15. (Continued)

- **Prime**

    1. **Primelink Software**—file transfer, virtual terminal, virtual printing, Prime PT2000 terminal emulation

- Tandem Computers, Inc.
- Merlo Business Systems

    1. **MacMerlo**—Tandem 6520 and 653X terminal emulation and text/graphic file transfer
    2. **Foundation Graphic Toolbox**—allows Tandem Computer resident graphic database to use Macintosh as a workstation
    3. **MAX**—host-based file transfer between Tandem and Macintosh

- Unisys
- **JBM Electronics**
- **APC-Univac**—asynchronous protocol converter supports Unisys Uniscope synchronous protocol
- Wang Laboratories
- **DataViz**

    1. **MacLink Plus/VS**—Wang word processing/Macintosh document interchange and file transfer

- **Omnigate**

    1. **The AllegroServer**—Wang terminal emulation, Wang-to-Macintosh document interchange and file transfer

IV. **PROPOSED CONNECTIVITY**
- IBM
- **AST Research**

    1. **Mac-86, Mac-286**—MS-DOS coprocessor board for Macintosh II and SE. Will let Macintosh run IBM Personal Computer XT or AT software

- **Digital Communications Associates**

    1. **Macirma SE, Macirma II**—3278 internal expansion card for Macintosh SE and II. Supports IRMA, Forte, and IBM Personal Computer 3270 host software

---

nications architectures, such as Wang's Systems Networking, IBM's SNA, and Digital Equipment Corp.'s DECnet. To provide the customer with multivendor integration, Wang provides products that are compatible with de facto standards such as SNA and TCP/IP.

In spite of its word-processing image, Wang has gotten behind the standards effort and was one of the first 18 vendors in the Computer and Communications Industry Association that formed the Corporation for Open Systems (COS). Wang heads Group 2 of COS, which is dedicated to designing reference implementations, or testing procedures, for assuring conformity among vendors' implementations of standards.

The Wang Systems Networking architecture for communications processors and systems, like Wang VS, is designed to support multiple protocols in an open-system environment. It allows the user to access and control layers of OSI protocols via a programming interface. The architecture allows Wang to build in support for new standards as required.

Wang is working on providing multiple local-area network bridging for 802.3 and 802.5. A phased implementation of EtherNet support on the VS level is also planned. For the support of local-area network standards, Wang has transceivers that allow 802.3 to run across a broadband Wangnet local net. For token-ring connectivity, products for IBM's Token-Ring local network will be available.

Wangnet is like 802.3 because it is based on a broadband version of the carrier-sense multiple access with collision-detection protocol.

Wang worked with the Institute of Electrical and Electronics Engineers to get its own proprietary protocol for local network broadband management and allocation accepted as a part of the 802.7 standard. The IEEE adopted it with a few minor modifications.

Wang has not committed to 802.4 and plans to provide connectivity from its systems to 802.4 through third-party board products.

The Technical and Office Protocol is supported by Wang as the office automation standard. A company with Wang products can tie terminals to VS with major LANs such as EtherNet, IBM's Token-Ring, and AT&T's Premises Distribution Systems.

Wang has a packet assembler/disassembler that allows a Wang workstation to last like an asynchronous device as it communicates with a remote host over X.25.

X.25 products also include a personal computer implementation that is used on WSN to transmit Wang protocols across X.25. Wang has a packet-layer interface that permits users to write their own implementations of a network application on top of X.25.

Wang has a public X.25 network called WangPac that customers can use as a backbone net. Wang offers support for X.3, X.28, and X.29 parameters on the VS, OIS, and at the PC level. The only limitation is that you cannot do word processing over an X.25 link, but for most applications that is not a real limitation.

FTAM, which is the OSI international standard for file transfer, has an early version with limited functionality in comparison to the second implementation. Wang's first implementation of FTAM has been demonstrated over 802.3, and X.25 and has been certified on OSInet.

In OSInet Wang demonstrated an FTAM prototype over OSInet that connected to other vendors. It also provided one of two X.25 networks for the project. This was its WangPac public packet-switching network.

Wang supports IBM protocols, including Binary Synchronous Communications, 3270, 2780, 3780, LU 6.2, DISOSS, and Professional Office System (PROFS).

Wang also supports other protocol emulations, including Burroughs Corp., Sperry Univac Minicomputer Operations, and VT100. It will provide general asynchronous interfaces to customers who wish to write their own asynchronous communications.

Wang works with third-party vendors and customers to provide interfaces so that users can write whatever protocol support they wish onto Wang products. For example, the X.25 packet layer protocol allows users or vendors to write TCP/IP onto VS. Wang has gateways to IBM, which piggyback on SNA transport with Wangnet protocols. They also have DISOSS and PROFS interfaces.

The Wang Office/DISOSS bridge product is not that easy to use because of the differences between IBM's and Wang's directories. Wang's protocol support includes bisyn and SNA for the VS as well as LU 6.2. For each protocol, however, you will need another physical line or port. There is a need to merge those capabilities so it can be done on one line.

OSI LIMITATIONS

OSI is not the only open-systems standard. The Consultative Committee on International Telephony and Telegraphy, the American National Standards Institute, and other standards organizations also develop standards but they coordinate their activities closely with one another and with ISO.

There are corresponding ISO standards for many CCITT developed standards, including the X.400 electronic-messaging protocols, which allow user-defined messages to be placed in a standard "envelope" and delivered to any electronic mail destination.

The original OSI architecture did not provide standard protocols for all functions necessary in multivendor networks. It did not address network management, so ISO developed separate network-management standards. OSI does not address specific user applications at layer 7, so additional application-specific standards are needed to link different vendors' spreadsheets, word processors, and other applications.

ANSDI's X12 EDI protocol illustrates the standards-within-standards situation that users must face. EDI application messages that conform to X12 will be carried by E-mail systems based on X.400. These will, in turn, be accommodated within layers 6 and 7 of an OSI-based communications architecture implemented in a gateway between proprietary network architectures.

X.400 and the CCITT X.25 packet-switching protocols are also accommodated within OSI. However, instead of cooperating to ensure that their implementations of these protocols match, different vendors implement these standards in different ways. Users that buy standard products from different vendors might find them to be incompatible.

A vendor might have an X.400 product that includes message-transfer agent capability but has no directory function, or the opposite can be true.

Because fully functional OSI-compatible products have not been available, many users are opting for interim alternatives. One of the best vendor-independent communications architectures that is available now is the Transmission Control Protocol/Internet Protocol protocol suite developed by the U.S. Department of Defense.

TCP/IP is the same concept as OSI, but it does not have the international backing OSI has. The same fundamental principle of providing generic communications functions that are not tied to any specific type of device or configuration applies to all of the different networking technologies.

At some point users must implement OSI rather than either TCP/IP or a mix of proprietary gateways to link vendor-specific architectures. TCP/IP is a good choice for users who have a variety of equipment in place already and need to tie it together now. Users that are entering into new long-term network projects or that do not need multivendor connectivity for their existing systems now might be better off waiting and then starting a full-scale OSI effort.

As fully functional OSI products become available, users that have implemented TCP/IP will probably start migrating to OSI. Migrations are difficult because you cannot do it layer by layer.

7

New Products

This chapter covers products that are important to network designers and managers. Some of these product areas have been with us for some time, but are now being driven by new lower cost manufacturing technologies, changing standards, and deregulation policies. This chapter is closely related to chapter 8, which treats emerging technologies and industries. One of the more important supporting product areas belongs to gateways and bridges. Another important product area is protocol converters.

GATEWAYS AND BRIDGES

Gateways and bridges are used to link local and remote networks. Many gateways are intended for local-area nets and hardware and software that is used with PCs.

Gateways allow products with dissimilar communications architectures to exchange information via multilevel protocol conversion, while *bridges* provide the logical links between physically distinct local-area networks that use identical protocols.

Some confusion exists as to what differentiates bridges and gateways. This stems from vendor advertising about using protocol converters and routers as bridges or gateways.

A *router* differs from a bridge or a gateway because it provides some services of both. A router operates at the network layer (OSI and Systems Network Architecture layer 3) and routes packets between multiple networks that use different protocols. Routers, like bridges, must determine where the packets are to be sent, and this causes packet-forwarding delays. Routers perform the packet segmentation and reassembly needed to accommodate intermediate networks in which packet sizes are different. Routers can transmit selectively packets based on urgency and allowable transmission delay.

A bridge is used to connect distinct physical networks into one logical network. All devices can address all others unless user restrictions are employed.

A bridge can act as a point-to-point store-and-forward device that accepts packets, checks their destination address in an internal table, and forwards packets addressed to another local network. Packets are filtered according to destination address and user-defined parameters.

Bridges also can be used to connect multiple local networks. The bridge on the output side of each network checks the address of each originating packet. If the destination is a workstation on another network, the packet is passed to the next logical network. Each bridge has to store only the destination addresses of the workstations on its network rather than all addresses in the logical network.

A local network bridge, operating at Open Systems Interconnect (OSI) layer 2 will link products that use identical protocols. The packet frame size must be the same, or the packets cannot be handled properly.

Smart bridges dynamically can record the location of the different stations rather than require the user to program the locations. Smart bridges allow devices to be added or deleted as the bridge routing tables are automatically updated.

Users needing to link EtherNet local nets should evaluate the EtherNet version supported by a bridge, because some versions cannot communicate with each other.

The media access control specification of IEEE 802.3 EtherNet uses a two-byte field length while EtherNet Versions 1 and 2 use the same address location for the type of field. The 802.3 version thus cannot communicate with Versions 1 and 2. Both types can exist on the same network and both versions in use can link each version to other similar networks but not to each other.

In a bridge, address checking and other internal functions can delay the forwarding of packets, reducing the volume of packets handled over a time period. This appears as the difference between the Packet/sec filtered and Packet/sec forwarded numbers show in TABLE 7-1. Most bridges are fast enough not to delay packets to the point at which applications will suffer. Some time-sensitive applications and those cases in which bridges are joining remote local networks may be sensitive to packet throughput.

The packet filter rate shown in TABLE 7-1 is based on an EtherNet packet size of 763 bytes and a token-ring packet size of 1 kbyte.

Gateways are more sophisticated because they support all seven layers of a network architecture, such as OSI or SNA. Gateways link networks with dissimilar protocols.

A gateway that is full-function means one in which an operation can be initiated at either end of the network. An IBM 3270, for example, might only be able to initiate a session with a host. These gateways that only emulate the 3270 are not full function.

Emulation and file-transfer capabilities do not make a full-function gateway.

Table 7-1. Typical Bridges.

Vendor/Product	Connection	Packet/Sec. Forwarded/Filtered	Filter Restrictions	Statistics Provided
Bridge Communications IB/X Series	EtherNet to EtherNet or broadband local network	8000/5000	User specified restrictions	Total forwarded, bad packets
Communication Machinery DRN-3100	Transmission Control Protocol/Internet Protocol EtherNet local networks	1000/300	Destination address	Total forwarded, bad packets
Digital Equipment Corp. LAN Bridge 100	Local or remote EtherNet Version 2.0 to IEEE 802.3	24,272/13,404	Destination address	Total passed, bad packets
Fast Feedback Technologies NetLink	Local or remote Fast Feedback MicroLAN	1000/500	Destination address	Total forwarded, bad packets
Netways Bridge Plus	Local or remote IEEE 802.3 EtherNet local network	10,000/5000	Destination address	Total forwarded, bad packets
TRW NB 2000	IEEE 802.3 EtherNet	10,000/5000	Destination address	Total forwarded, bad packets
Vitalink Communications TransLAN	EtherNet local networks	15,000/2500	Destination address, packet type, multicast address, user-specified code	Total forwarded, bad packets

Reverse pass-through, which is the ability to initiate a call in either direction, is needed. This requires a host-command facility.

Some gateway vendors do not provide suitable support with their products. LU 6.2 applications can require users to write their own Application Program Interface (API), which requires a knowledge of LU 6.2 and API procedures. End users might not be familiar with either of these.

The LU 6.2 verbs can be powerful, but the application needs to be designed so that it can manipulate them. Some vendors provide this support for their customers while others do not.

Gateway selection is even more complex because of the different levels of services provided. Some gateway vendors claim that the protocol conversion they use to link dissimilar networks is transparent to the user. This is not necessarily so, especially if file transfers are being performed.

Simple gateways that are little more than protocol converters usually require the least user involvement, while those employing LU 6.2 can require programming services that many users might not have in-house.

Even something as simple as IBM 3270 emulation can require user-supplied programming, and the data-processing staff might be required to write routines that convert host data into a form personal-computer applications can use. This is especially true for the transfers, because data conversion (ASCII to and from EBCDIC) and file blocking and unblocking usually are involved. However, straightforward interactive applications such as those involving a database can use a standard API that is not hard to write. Some LU 6.2 vendors might have those interfaces available. More specialized applications such as user-written accounting and order-tracking packages might require a good deal of custom development.

In spite of the programming effort required, gateways are still the easiest way to link diverse networks. Vendors and consultants can help out with the API, provided they have had experience with similar applications. In the future, some of the more common applications, such as databases, are likely to have standard APIs that are easily implemented.

The widespread acceptance of gateways has been hampered by the lack of a single network monitor, although NetView is rapidly becoming the needed standard.

Although there might be no simple implementation of gateways and protocol converters, bridges and routers can be installed with little or no user programming. In TABLE 7-2, routers are listed with gateway products, but they do not support protocol conversion or provide support above the OSI network layer and SNA path-control layer. Routers can be identified by their OSI layer-3 support designation under **Layer** support in TABLE 7-2.

PROTOCOL CONVERTERS

The common method of providing effective, low-cost communications for a diverse mix of mainframes and terminals is to give each terminal its own com-

Table 7-2. Gateways. (Continued thru page 219)

Vendor/ Product	Connectivity	Emulation	Protocols/ Simultaneous Sessions	Layer of Support	Trunk Speed (Bit/Sec)
Data General DG/TCP IP gateway	EtherNet TCP/IP local network to any local network running TCP/IP	N/A	TCP/IF X.25 level 3 N/A	OSI 7	19.2k packet networks 10M local network connection
DG/SNA	DG MV minicomputer to IBM SNA	IBM PU 2, LU 0,1,2,3, 6.2	SDLC, X.25 level 3 254	SNA 7 wide-area networks	19.2k packet networks; 56k
Datapoint Vista-gate	Any Datapoint local network to SNA	IBM PU 2.1, LU 1, 2, 3, 6.2	SDLC, X.25 level 3 30	SNA 6	64k
Digital Communications Associates IrmaLAN to SDLC gateway	IBM Personal Computer to SNA	IBM 3274/3278, 3279	SDLC 32	SNA 7	19.2k
IrmaLAN DFT gateway	IBM-compatible NETBIOS via EtherNet to SNA	IBM 3274/ 3278, 3279	SDLC 20	SNA 7	10M (EtherNet)
Digital Equipment Corp. DECnet to SNA gateway	DEC VAX to SNA	IBM PU 2m LU 0,1,2,3,6.2	SDLC, SNADS N/A	SNA 7	56k
DSC Nestar Systems Asynchronous Server versions 2.0	Nestar Plan local network to any asynchronous host	Asynchronous terminal	Asynchronous 16	OSI 7	19.2k

Vendor/ Product	Connectivity	Emulation	Protocols/ Simultaneous Sessions	Layer of Support	Trunk Speed (Bit/Sec)
SNA gateway version 3.0	Nestar Plan local network to SNA	IBM 3274/3278, 3279	SDLC 16	SNA 7	19.2k
Fox Research 10-Gate Turbo	TenNet local network to SNA	IBM 3274/3278, 3279, 3286, 3287 3289	SDLC 32	SNA 7	19.2k
Novell, Novell NetWare Remote Bridge X.25	Any NetWare-controlled local network to same	N/A	X.25 32 or 256	OSI 4	64k
Orion Network Systems SNA 6.2 Release 3 gateway	Any workstation to SNA	Any IBM PU 2.1, LU 0,1,2, or 6.4	SDLC N/L	SNA 7	56k
Proteon Model 4200	Any IEEE 802.3 EtherNet, ProNet 10 or 80 or ARPANET 1822 to each other	N/A	EtherNet N/A	OSI 3	2.048M
Rabbit Software Rabbitgate	Any remote IBM NETBIOS local network to SNA	IBM 3274, 3276/3278 3279, 3287, 3770	SDLC, BSC, SNA/X.25 level 3 32	SNA 7	19.2k packet networks; 56k wide-area networks
3Com 3 + /asynchronous	3Com or IBM Token-Ring, 3Com Ethernet to any asynchronous host	IBM 3101, DEC VT 52, 100, 220	asynchronous 2	OSI 7	19.2k
Software Results Comboard/SNA	DEC VAX or Micro VAX to SNA via EtherNet, direct connect to host or remote connection	IBM 3274/ 3278, 3287, 2777	SDLC 30	SNA 7	56k

Comboard/HASP	DEC VAX or Micro VAX to IBM BSC via EtherNet, direct connect to host or remote connection	IBM HASP workstation	BSC 255	SNA 2	56k
Token Associates MAP/X.25	MAP broadband or baseband local network to X.25 or to another	N/A	MAP X.25 layer 3 32	OSI 7	56k
Tri-Data Netway 1000A	Apple AppleTalk to SNA/SDLC or BSC	IBM 3274/3278	SDLC, BSC 16	SNA 5	19.2k
Ungermann-Bass Net/One X.25	Ungermann-Bass 802.3 Net/One to X.25	N/A	X.25 level 3/32	OSI 5	64k
Wang Laboratories SNA Networks	Wang VS super-minicomputer to SNA	IBM 3274/ 3278, 3279 3777	SNA/SDLC N/L	SNA 7	56k
DISOSS	Wang Office to IBM DISOSS	N/A	LU 6.2 N/L	SNA 7	19.2k

Key:
APPC—Advanced Program-to-Program Communications
ARPANET—Advanced Research Projects Agency Network
BSC—Binary Synchronous Communications
MAP—Manufacturing Automation Protocol
N/A—not applicable
NETBIOS—Network Basic I/O System
N/L—not limited
OSI—Open-Systems Interconnect
SDLC—synchronous data-link control
SNA—Systems Network Architecture
SNADS—System's Network Architecture Distribution Services
TCP/IP—Transmission Control/Internet Protocol
XNS—Xerox Network System

219

munications link or to employ protocol converters. Either method can be an expensive proposition. Some products require extensive programming just to support file transfers.

Protocol converters are popular because low-cost asynchronous terminals can be used to emulate IBM 327X Model 2 workstations. Protocol converters used with PCs usually consist of expansion boards with software.

Other stand-alone processors are tabletop units. They handle all networking functions, provide host processor communications, and perform all communications switching between terminals and host processors. A disk/diskette is used for storing network configuration information, terminal personalities, protocols, and a Nodal Processor operating system.

The Nodal Processor (NP) is the central component in the network and can be offered in several versions, the differences between them being the size of the built-in hard disk and/or floppy disk used.

A typical Nodal Processor will contain an 8-MHz processor, 512 kbytes of memory, and four serial data communications ports for host processor connections or for linking Hubs or other NPs. Host communications speeds might be 19.2 kbit/s asynchronous or 56 kbit/s synchronous.

Some units support internodal leased-line, or X.25 connections and can be used as gateways or X.25 networks as well as for token-passing local-area networks.

A single terminal can converse simultaneously with multiple hosts. Terminals can either be locally attached to the converter, or else users can dial into them.

They differ in system architecture, the number of simultaneous applications permitted per user, and protocol-conversion services provided. Some converters use a physically distributed architecture in which terminal devices interface with stand-alone terminal handlers called *device interface processors*. These furnish terminal management and protocol-conversion services. Their output is fed into the Nodal Processor, which provides networking functions, switching, and host-processor interaction.

Other systems use a central processor unit (a data communications computer) to handle terminal interfacing, protocol conversion, and terminal and host switching. The actual protocol conversion services are distributed within the central processor and are shared by all terminals.

The major difference between the interface processors is in the types of attachable devices. Some provide a coaxial-A interface for IBM terminals and printers, while others furnish an RS-232-C interface for terminals, printers, and point-to-point hosts.

An interface processor can be configured to work with several protocol converter types such as IBM Systems Network Architecture/Synchronous Data-Link Control, Binary Synchronous Communications, Honeywell's VIP, Burroughs' Poll-Select, Sperry Univac's UTS, Digital Equipment Corp. and Data General asynchronous, Hewlett-Packard's ENQ-ACK, IPARS, SABRE, and

X.25 Level II, and X.25 BSC. The protocol suites usually are not standard equipment, and each must be purchased separately.

Terminal keyboards usually are mapped to permit function keys to operate in native mode. The exception is where the terminal being emulated has no corresponding key. A dumb ASCII terminal emulating an IBM 3278, for example, can require multiple keystrokes to emulate a function normally activated through a programmable function key.

Most protocol converters let customers change the keyboard map to suit their applications, but some are easier to use than others. Any terminal can be connected to establish concurrent sessions, with sessions from different hosts displayed concurrently through split screens. Users can manipulate data on the screen and hotkey between sessions. There must be sufficient buffer storage for handling inputs from the host processor, even when the addressed session is suspended and not active on the screen. When a suspended session is activated, this allows the data to be displayed automatically.

Software products are available for file transfers. They usually are table-driven and use a combination dictionary/directory to identify and define record layouts and the form the data must be in to use the application. They also change EBCDIC to ASCII and convert host files to a form personal computers can use.

While file-transfer products are important for protocol converters, user training is another important issue, especially if IBM 3270 emulation is involved. Operators with no 3270 experience will need training to recognize and react to system messages so they will not insert commands that confuse the host.

Communications processor boards also are offered for linking EtherNet IEEE 803.3 with an EtherNet IOP to form a network bridge. IOPs can be linked through an interval bus to share services among the communications-interface boards. Thus, users with multiple types of terminals at one location can use one interface board to have several types of protocol services performed. The link can be handled by a system administrator using a network-definition language.

Other systems allow a single protocol converter or an entire suite to be changed dynamically to meet application needs by downloading the required software.

Most converter products (TABLE 7-3) are close to even in terms of services furnished, but some systems require less hardware. Protocol-sharing systems could offer an edge in that everything can be generated at system initialization time, as opposed to dynamic downloading. A lot of activity in a protocol module could cause performance problems, however, unless multiple versions of the same protocol are generated.

Systems are available that support peer-to-peer communications using APPC and SDLC between a wide range of IBM systems, including System/36 and /38, Series 1, 5520, Personal Computer XT and AT, 9370 with VM/APPC,

Table 7-3. Typical Protocol Converter. (Continued thru page 227)

Vendor Model	Package	Host Interfaced	Micro-computer Supported	Host Communications Software	Micro-computer Memory Required	Terminals & Printers Emulated	Transmission Speed & Protocols
Adacom CP-301	H & S	IBM 3X74	IBM Personal Computer	N/A	128 kbytes	3278 proprietary	19.2 kbps
AST Research AST-SNA	H & S	IBM System/ 36, 303X, 43XX	IBM Personal Computer	MVS/TSO, VM/ CMS	192 kbytes	3278, 3279; 3287	9.6 kbps; BSC, SDLC
Avatar Mac Main Frame	H & S	IBM 30XX, Inc. Macintosh 3X74 controller	Apple Computer with IBM	Interfaces	Not available	3278	Depends on 3X74
CXI PCOX/Plus Remote	H & S	IBM 30XX 43XX, 9370	IBM Personal Computer	VM/CMS, MVS/ TSO, CICS/VS, DOS/VSE	340 kbytes	3278, 3279; 3287	19.2 kbps; SDLC
Cincom Systems PC Contact	S	IBM 30XX, 43XX	IBM Personal Computer	VM/CMS, MVS/ TSO	128 bytes	3278, 3279	19.2 kbps; SDLC
Cleo Software	H & S	IBM 30XX, 43 XX	IBM Personal Computer	MVS/TSO, VM/ CMS, CICS/ VS	128k or 256k bytes	3278 Cleo-3270	9.6 kbps; SNA or BSC, Cleo = 5250 SNA
Communications Research Group Blast	S	IBM DEC Data General AT&T, Hewlett Packard	IBM Personal Computer AT&T, Apple II Macintosh	VM/CMS, MVS/ TSO (IBM); VAX VMS, PDP/ RSX, RT-11, (C_O); AOS, AOS/VS (DG); 3000/ (HP) MPE, 1000/RTE	128 kbytes (IBM; 512 kbytes (Apple)	DEC VT100 DG D-200	4.8 kbps asynchronous; 19.2 kbps direct connect
Communications Solutions Access SNA	S	IBM 30XX, 43XX	IBM Personal Computer	MVS/TSO, VM CMS, CICS/VS, IMS/VS	192 bytes	3278, 3279 3287	9.6 kbps; SDLC
Computer Corp. PC-204	S	IBM 30XX 43XX	IBM Personal Computer	MVS/TSO,/ CICS/VS	384 kbytes	3278, 3279	9.6 kbps; BSC

Product	H/S	Host Computer	PC	Operating System	Memory	Terminal Emulation	Speed/Protocol
Computer Vectors RCOM 2	S	IBM 30XX 43XX	IBM Personal Computer	MVS/TSO, CMS, CICS/VS DOS/VSE	256 kbytes	3277, 3278, 3279	19.2 kbps; asynchronous
Cullinet Software Infogate	S	IBM 303X, 43XX	IBM Personal Computer	MVS/TSO, VM/CMS, CICS/VS,	384 kbytes	3278, 3279	19.2 kbps; BSC, SDLC
Datagram E. Greenwich, R.I. Model 24XX	H & S	Burroughs	Personal Computer	Any Burroughs	Not available	Any Burroughs	19.2 kbps; asynchronous
Digital Communications Assoc. IRMA Comm	H & S	IBM 303X, 43XX	IBM Personal Computer	MVS/TSO, VM/CMS, CICS/VS	320 bytes	3278, 3279	9.6 kbps; SDLC
Digital Equip. Corp. DECnet Dos	S	DEC VAX, Micro VAX, IBM 30XX, 43XX	IBM Personal Computer	DECnet, VM CMS MVS/TSO,/CICS/VS	256 bytes	DEC VT 100/200, IBM 3278	9.6 kbps; SDLC, BSC
Diversified Resources RCOM PC	S	IBM 30XX 43XX	IBM Personal Computer	VM/CMS,/CICS/VS	128 bytes	3278	19.2 kbps; SDLC
Dynatech Packet Tech. CPX-25	H & S	Any supporting X.25	Personal Computer	Any supporting X.25 or PAD	N/A	Asynchronous	19.2 kbps; X.25, X.3, X.28, X.29
East Com. Syncra	S	IBM System/370, 30XX, 43XX, System/3; DEC VAX, POP-11	IBM Personal Computer	VTAM	256 kbytes	3278, 3279 VT100	Up to 56 kbps; direct attach
Hayes Smartcom II	S	Any DEC, Macintosh	IBM Personal Computer, Apple	VAX/VMS	192 kbytes (IBM); 128k	DEC VT52, 10X	9.6 kbps; asynchronous
ICOT X.25 PC Adapter	H & S	IBM 30XX, 43XX	IBM Personal Computer	MVS/TOS, VM CMS, VM, CICS/VSVS	56 kbytes	3278, 3279	19.2 kbps; X.25 & SNA/HDLC
Integrated Network SDLC PC	H & S	IBM 30XX, 43XX	IBM Personal Computer	MVS/TSO, VM CMS, CICS/VS, DOS/VSE	55 kbytes	3278, 3279 3287	19.2 kbps; SDLC

Vendor Model	Package	Host Interfaced	Micro-computer Supported	Host Communications Software	Micro-computer Memory Required	Terminals & Printers Emulated	Transmission Speed & Protocols
JDS Micro-processing Assoc. Hydra II	H & S	IBM System/360, 370, 30XX, 43XX	IBM Personal Computer	MVS/TSO, VM/CMS, CICS/VS	Varies with 3-party vendor; works with VM Relay, Master Link, & RCOM	3278, 3279	Hydra II board interfaces with byte multiplexer channel; 19.2 kbps; asynchronous
KMW Systems Series II Twinax	H & S	IBM System/3X	IBM Personal Computer, Apple Macintosh, Wang Labcratories Personal computer	N/A	N/A	IBM 5291	19.2 kbps; asynchronous
Local Data DataLynx/3174	H & S	IBM 30XX, 43XX	IBM Personal Computer,	VM/CMS, MVS/TSO, CICS/VS	128 kbytes	3278, 3279; 3287	19.2 kbps; SDLC asynchronous
Management Science America ExpertLink	S	IBM 30XX, 43XX	IBM Personal Computer	CICS/VS, IDMS/DC, TPMON	256 kbytes	3278	9.6 kbps; BSC, SDLC
McCormack & Dodge Millenium PC	S	IBM 30XX, 43XX, 8100	IBM Personal Computer	CICS/VS, IDMS	256 kbytes	3278	9.6 kbps; BSC, SDLC, asynchronous
Micom Systems M74XX	H & S	IBM 30XX, 43XX	IBM Personal Computer	VM/CMS, MVS TSO CICS/VS, DOS/VSE	126 kbytes	3278	19.2 kbps; BSC, SDLC, asynchronous
Micro Tempus/Tempus Link	S	IBM 30XX, 43XX	IBM Personal Computer	MVS/TSO, VM/CMS, CICS/VS	128 kbytes	3278	19.2 kbps; BSC, SDLC X.25
Multi-Soft Super-Link	S	IBM 30XX, 43XX; DEC VAX	IBM Personal Computer	MVS/TSO, VM/CMS, CICS/VS	384 kbytes	3278, 3270; TTY	10.2 kbps; SDLC or BSC
On-Line Business Systems	S	IBM 30XX, 43XX	IBM Personal Computer	VM/CMS, MVS/TSO	72 kbytes	IBM 3101	9.6 kbps; asynchronous

Product	S / H&S	Hardware	Computer	Software/OS	Memory	Terminal	Protocol/Speed
On-Line Software Intl. Omnlink	S	IBM 30XX, 43XX	IBM Personal Computer	CICS/VS	256 kbytes	4.8 kbps; 3278	SDLC, asynchronous
Orion Group SNA LU 6.2	S	IBM 30XX, 43XX, System/36 & 38	IBM Personal Computer	CICS.VS	640 kbytes	Any LU 6.2/ PU 2.1	56 kbps; SDLC
Pathway Design NetPath 3270	H & S	IBM 30XX, 43XX 8100 Series I	IBM Personal Computer	VM/CMS, MVS/TSO CICS/VS VTAM, BTAM	256 kbytes	3277, 3278, 3279, 3284, 3286	19.2 kbps; SDLC
Performance Software Master-Link	S	IBM 30XX, 43XX	IBM Personal Computer	MVS,TSO, VM/CMS, CICS/VS DOS/VSE	256 kbytes	None. Needs 3-party board such as Irma	19.2 kbps; asynchronous, synchronous, synchronous, SD_C
Perle Systems PDS-350/294	H & S	IBM System/3X	IBM Personal Computer	IMS/VS, CICS/VS	128 kbytes	525k	19.2 kbps; SDLC
Persoft SmartTerm-240	S	DEC VAX	IBM Personal Computer	VAX/VMS	512 kbytes	DEC VT52, 100,200	
Protocol Computers Model 276	H & S	IBM System/ 370, 303X, 43XX	IBM Personal Computer	VM/CMS, MVS/TSO CICS/VS	128 kbytes	3278-2; 4387	19.2 kbps; SDLC
Remex TMS-1	H & S	IBM System/ 370, 30XX, 43XX	IBM Personal Computer	VM/CMS, TSO	512 kbytes	3278, 3279, 3862	19.2 kbps; BSC, SDLC
Simware Sim-PC	S	IBM 30XX, 43XX, System 370	IBM	MVS/TSO, VM/ CMS, CICS/VS, VTAM	192k- or 384 kbytes	3278, DEC VT100; 3287	19.2 kbps; SDLC, BSC, asynchronous
Sterling Software, Micro/Answer	S	IBM 30XX, 43XX	IBM Personal Computer	CICS/VS, MVS/TSO	256 kbytes	3278	9.6 kbps; SDLC

Vendor Model	Package	Host Interfaced	Micro-computer Supported	Host Communications Software	Micro-computer Memory Required	Terminals & Printers Emulated	Transmission Speed & Protocols
The Systems Center	S	IBM 30XX, 43XX	IBM Personal Computer	DOS/VSE VTAM	300 kbytes	None. Needs 3-party board such as Irma	9.6 kbps; SDLC
Tangram Systems/ Arbiter	S	IBM 30XX 43XX	IBM Personal Computer	VTAM	32 kbytes	3278	56 kbps; SDLC, BSC
Techland Systems/Bluelink 3270	H & S	IBM 30XX 43XX	IBM Personal Computer	MVS/TSO, VM/CMS	23 kbytes	3278	4.8 kbps; SDLC
Thomas Eng.	H & S	Honeywell,	Any	DSA (Honeywell),	256 kbytes	Honeywell,	19.2 kbps;
Uni-Tec		(all), IBM System/370, 30XX, 43XX, Unisys Corp. 1100	Personal Computer	CMS/TSO (IBM), CMS 1100 (Unisys)		IBM 3278, Unisys UNISCOPE	BSC, SDLC (IBM); HDLC (Honeywell); CMS 1100 & Telcon (Unisys)
Universal Data Systems Sync-Up SNA/3270	H & S	IBM 30XX, 43XX	IBM Personal Computer	VM/CMS, MVS/ TSO, CICS/VS	256 kbytes	3278, 3279, 3287, 3289	9.6 kbps; SDLC
VM Personal Computing Relay Gold	S	IBM 30XX, 43XX	IBM Personal Computer	MVS/TSO, VM/CMS	192 kbytes	3278	19.2 kbps; proprietary
Winterhalter Data Talker 3270	H & S	IBM 30XX, 43XX	IBM Personal Computer	MVS/TSO, VM/CMS	128 kbytes	3278, 3279	19.2 kbps; BSC

Key:
ALT—alternate
APL—A Programming Language
BSC—Binary Synchronous Communications
CMS conversational
DC—data communications
HDLC—high-level data-link control
NA—not available
PAD—packet assembler/disassembler
SDLC—synchronous data-link control
SNA—Systems Network Architecture
TPMON—TP monitor
TSO— me sharing option
VSE—virtual storage
WS—windows supported

Displaywriter, and Scanmaster. LU 6.2 also is implemented on System/370, 30XX, 43XX, and 8100 using PU 5.0 communications. Systems are offered that implement the IBM SNA extension to let users transfer files between the System/370, System/38, personal computers, and Digital Equipment Corp. VAX/VMS machines.

Even with the flexibility and availability of LU 6.2, protocol converters are still a good buy for now, but users should begin planning for LU 6.2 because it operates in native mode and allows peer-to-peer communications.

CALL-MANAGEMENT SYSTEMS

A call management system can

- Monitor telephone call traffic
- Point out peak periods and the number of units needed to handle different call volumes
- Identify factors, such as efficiency, that adversely affect the quality of the call-answering service

Typical of such products are the MIS-1 from Telecalc and the RT-10 from Perimeter Technology. Both are personal-computer-based software products that attempt to solve the most common phone-room problems. They provide essentially similar services. The Perimeter product works with both private-branch exchanges and Centrex, while the Telecalc unit is for Centrex only.

Another call monitoring system is Telco Research's TRU telecommunications-tracking system, which uses an IBM Personal Computer XT or AT to monitor and analyze each workstation's call traffic as it passes through the PBX.

The use of call-monitoring systems to ensure a high degree of productivity is growing. Rather than just listening to telephone conversations, they perform detailed analysis of employee phone usage over time. More than a third of the 20 million Americans who work at computer terminals and telephones are currently monitored.

By the year 2000, 60 to 70 percent of the estimated 50 million employees who use computer terminals and telephones might be monitored.

A typical call monitoring, or accounting, system can cost from $6000 to more than $100,000, depending on the number of telephones and the user's communications architecture. Over 200 companies currently sell call-monitoring devices and software, and many of these are new, unestablished firms.

Even the better systems can fall behind in the rapidly changing telecommunications industry.

PBX formats can change and the database of tariffs and communications providers must be updated. In this volume market, call-accounting systems also must change if they are to continue to function over time. Maintenance and support is important as well as the types of reports the system compiles over time. This can include detailed usage data for each employee, including the time and data of every call, the numbers called, call duration and cost, as well as a break-down of exception calls (calls that are not job-related or exceed a specified dura-tion or cost). Other more generic call traffic data is also generated, such as the number and duration of incoming and outgoing calls each hour at every worksta-tion. The result is a telecommunications version of an audit trail that can depict the activities of every employee.

Call-monitoring systems and the data they generate can provide an accurate portrait of a company and its employees that cannot be obtained through any other means.

Communications is a reflection of a business, just like a financial statement. By understanding the different kinds of information generated by a call-accounting system, a company can learn why certain functions are successful.

Suppose a supplier frequently delivers faulty items to a manufacturer, but not often enough for anyone to notice. Monitoring telephone usage with a call-accounting system could detect this problem because of the sudden increase in telephone calls from the manufacturing firm's service department to the sup-plier.

The call-management system can spot a quality-control problem before supervisory personnel notice it. Systems like the Telecalc, MIS-1, and Perime-ter RT-10 provide real-time status information for phone-room management. Most systems monitor levels of call activity, showing the status of trunks, calls in progress, calls waiting in queues, time before calls are answered, length of calls, calls lost due to hang-ups and agent status. Those using the systems receive information regarding agent performance, such as the number of calls handled, length of each call, percentage of time spent on the telephone, average time on hold, and time spent waiting for calls. Information is also provided showing the performance of agent groups and departmental productivity.

Most systems compile these performance statistics into historical files that are used to calculate and forecast staffing schedules to meet specified perform-ance levels. Some even generate an economics report that determines the most profitable level of service needed to operate.

By running the forecast and schedule data through an optimization pro-gram, you can produce a report showing the number of busy calls and hang-ups, which the customer can use to determine whether or not it's economically worthwhile to reschedule.

These systems typically gather call information through the use of a line scanner. The scanner resides at the customer's premises and bridges the telephone-switching equipment, Centrex or PBX, to monitor the activity on the trunk and station lines.

The line scanner's output is fed into a personal computer (XT or AT-level machine), and the computer does the computations needed to provide the man-agement information. This includes both real-time and printed information. Users can obtain summary statistics for agent or group performance, or the performance of individual agents can be presented. Typical displays or printouts include incoming and outgoing calls, average length of calls in minutes, total conversation time, and agent status (on phone, on hold, available, or unavailable) for any specified time of day. Users can also obtain performance information such as the number of rings before an agent picked up a call and the length of time callers waited in the queue before hanging up.

Displays of information can include color-coded bar graphs that show trunk-group activity and indicate the number of calls in a particular condition. Color bars and numeric designators can be provided for all trunks and attendant groups. Another technique is to show trunk status and agent activity in real-time, columnar form. Displays can show trunk status, showing the number of trunks handling incoming calls, outgoing calls, and trunks in queue or idle.

These systems provide a macro or micro view of system status, and includes such data as the average call-answer speed, maximum expected call-answer delay, holding time of the longest call, number of calls handled today, and the number of calls abandoned. This data can be used to determine threshold parameters and activate alarms when they are approached. Some systems might also be able to calculate trunk overflow projects from an analysis of the average length of time a caller spends in a queue before hanging up.

This time can be used to establish performance thresholds that will indicate when a large number of caller hang-ups will occur. Some systems are not able to perform overflow projections in real time because they do not accumulate time-in-queue and busy-line data. Such data must be obtained from the local telephone company central office.

The more flexible systems will function with any vendor's automatic call-distribution (ACD) facilities, as well as with uniform call-distribution (UCD) services associated with Centrex systems. Others can be more restricted, working only with Centrex and UCD.

Some will work with digital PBXs but will monitor only analog telephone calls. This is a decided disadvantage given the growth of the digital PBX market.

Besides presenting the data on the personal computer's monitor and allowing hard copies to be printed out, information presentation is offered in a wall-hanging, flat-panel display.

These units are well suited for large organizations that require regional offices to report phone-room activity to some central location. Using an *executive-polling* feature, a manager can poll up to 150 remote systems overnight to receive activity summary reports showing call-activity detail, call volume handled, and quality of service.

The graphics and color-coded situation-analysis features allow much useful information to be shown at one time. It also has the capability to present more real-time information. Real-time information is needed to manage telemarketing and service operations effectively. You can see exactly what is going on with all trunks and each agent on one screen.

Call-monitoring systems can provide an essential service by offering information that is only available through a studied analysis of the corporate telecommunications environment. When implemented without discretion, these systems can make communications management appear as a spy and collector of useless information. When used wisely, however, they can enhance communications management and help isolate and address problems confronting the corporation.

When used as a tool for feedback and not as a means of invading privacy, monitoring systems are perceived by employees as an educational device that helps them do their jobs better.

VOICE-MESSAGING SYSTEMS

Voice-messaging systems are available in many forms and at many levels of sophistication. Some provide automated attendant and voice sending and receiving only, while others add features such as networking, outcalling, message distribution, transaction processing, E-mail, and integration with PBX/Centrex facilities.

In addition to the basic services enabling users to send, receive, and copy messages, many systems act as automated call attendants that can supplement human private-branch exchange or Centrex operators.

Voice-messaging systems can also be used as dial-up information services. A bulletin-board application would let everyone call in for recorded information. Some companies are using voice-message systems to recruit employees from classified ads. Applicants phone in for details and those interested are invited to enter a certain code for more information to enter a different code. Another application is transaction processing, in which callers can enter orders for processing.

Voice-messaging for electronic-mail integration is another application. Systems supporting E-mail leave a flag in the mailbox indicating that a message is waiting. Some systems, such as VMX's, can convert text messages to voice.

Voice-messaging systems that interface with telephone lines or trunks usually are installed at the PBX. They use coders and decoders to convert analog voice to digital signals and store the quantized voice on disks.

They let the voice-messaging system invoke PBX services such as activating the message-waiting indicator. Complete PBX integration is difficult because vendors do not always publish their interface specifications.

Systems can be limited in invoking many PBX services. Most full-featured, voice-messaging systems are stand-alone units with their own hardware, software, and disk drives. Others operate on an IBM Personal Computer AT and XT disk systems and use special software for message processing. Personal computer-based systems are useful for about four ports.

Most systems provide an automated attendant that answers calls with a greeting and asks the caller to enter the desired extension number. The caller must have a pushbutton telephone.

Systems from Octel Communications, Rolm, AT&T, and Wang Laboratories allow callers to enter via pushbutton keypad all or part of the called party's last name. Based upon the tones entered, the system then provides a list of extensions.

The automated system usually handles voice mailbox services and prompts the caller to use the available system options. The system can prompt for identification codes, tell how to leave a message, review its contents, and provide services such as marking the message for urgent or normal delivery or calling another extension.

Urgent messages are presented to mailbox holders ahead of regular messages. Messages marked urgent also can be delivered on a priority basis rather than by normal batch delivery.

Users also can mark a message as private, which prevents the recipient from sending it on to other boxholders. Many systems let mailbox holders establish guest mailboxes within their own private boxes in order to accommodate nonsystem users. The guest box uses a password assigned by the primary voice box holder and lets the caller take and respond to messages to and from the primary box holder. The guest mailbox user can direct messages to other box holders, in addition to the primary box holder. Most systems allow messages to be sent automatically to other mailboxes without the sender entering every address. This facility uses distribution lists containing target mailbox addresses.

Most systems allow users to define their own lists. Some have restrictions on the number of lists per user and the number of names per list. Systems that support limited lists and names include those by Northern Telecom, Rolm, Votan, AT&E Centigram, and Miami Voice.

As part of the total message-delivery facility, many systems support outcalling to remote locations such as home telephones. The level of outcalling services varies, but most support delivery based on date, time of day, and urgency, and many support beeper-call services.

The establishment of voice mailboxes in which individuals send and receive messages is normally done by a system administrator who specifies the maximum message length and message capacity. The administrator might also specify an archive limitation to prevent users from occupying too much disk space. Many box holders tend to save everything, and disk storage needs can be abused.

Limits are needed generally on how long messages can be held. The storage needed depends on the quantization scheme used. Pulse-code modulation, for example, uses 64 kbps for word conversion. Voice-messaging systems usually specify quoted capacity of their systems in storage hours provided, not by bytes of disk capacity. The storage technique for words is different from that for data.

In TABLE 7-4, the number of ports supported by a voice-messaging system refers to the number of simultaneous inputs that can be accommodated. A

Table 7-4. Voice Messaging Systems. (Continued thru page 235)

Vendor Model	Ports	Storage Hours	Telephone Interfaces	Outcalling	Networking Nodes	Private-Branch Exchange	Electronic-Mail Integration
AT&T Audix	2-32	10-192	Ground, loop E&M, DID	B		1	AT&T E-Mail
American Telesystems Express Messenger	Up to 24	Up to 86	Loop	T,U,B	Unlimited	4,7,8,10	IBM PROFS, Digital Equip., Data Gen., Novell 3Com Corp.
Applied Tech. Call-xpress	Up to 8	Up to 13	Loop, DID	T		2,3,7,8	
BBL Indust-ries EVXTRA	Up to 32	8 to 60	Ground, loop E&M DID	T,U,B		8,10	
Brooktrout Tech., V-Mail 210	1 to 3	1.5 to 3.5	Ground, DID	T,U,B		1,6,8,10	
Centrigram Voice Memo	4 to 14	Up to 30	Ground, loop, E&M, DID	T,U,B		2,4,5,6,7, 8,10	
Comverse Tech. Trilogue	4 to 32	8 to 182	Ground, loop, E&M DID	T,U,B	Unlimited	1,2,6,7, 8,9,10	DEC VAX/VMS, Sperry Corp., Notes

System					
Digital Sound DSC-2000	2 to 20	6 to 130	Loop, E&M DID	T,U,B	40	1,2,4,6,8
Dyte/Automated Attendant Exchange	Up to 8	Up to 20.5	Ground, loop,	T,U,B		All
Innovative Tech., Receptionist	4 to 12	Depends on personal computer used	Loop, DID	T,U,B		1,6,7,9,11
InteCom InteMail	Up to 96	Up to 272	Ground start	T,U,B	Unlimited	
Message Processing Autotend/Voice Relay	4 to 24	3 to 100	Ground, loop DID	T,U,B	Unlimited	1,6,7,8,9
Miami Voice/Miami Voice	4 to 16	1.5 to 47.5	Ground, loop	T,U,B		11,12
Microvoice Sys. Aspex Automated Operator/Voice Mail	4 to 16	Up to 19	Loop, DID	T,U,B	2	
Northern Telecom/Meridian Mail	4 to 32	10 to 208	T-1, DTI			8

233

Vendor Model	Ports	Storage Hours	Telephone Interfaces	Outcalling	Networking Nodes	Private-Branch Exchange	Electronic-Mail Integration
Octel Communications Aspen	4 to 72	6 to 304	Ground, loop E&M, DID	T,U,B	500	1,2,3,4,5, 6,8,9,10, 12	
Perception Tech./Interactive Voice Response	8 to 48	More than 100	Ground loop, DID	T,U,B	Unlimited	All	
Periphonic TriLogue	4 to 32	8 to 70	Ground, loop, E&M, DID	T,U,B	Unlimited	1,2,8,10	DEC VAX/Mail
Rolm PhoneMail	4 to 128	5 to 480	Ground, loop, E&M, DID	B	50	2,9	IBM PROFS & VTMS
Voicemail Intl./Voicemail Information System	4 to 128	25 to 400	Ground, loop E&M	T,U,B	Unlimited	2,7,8	DEC All-In-1
VMX 5000 Series	12 to 64	26 to 526	Ground, loop, E&M, DID	T,U,B	Unlimited	1,2,3,4,5, 6,7,8,9	DEC VAX/Mail All-In-1 IBM PROFS
Votan/TeleCenter	1 to 4	4 to 15	Ground, loop				

Votrax Intl./Voice Message Center	2 to 48	Up to 1000	Loop	T,U,B		1,8,9	
Wang Laboratories/DVX	4 to 24	Up to 1344	Ground, loop		80	2,5,8	
Xerox/XVMX	4 to 64	12 to 526	Ground, loop, E&M, DID		Unlimited	1,2,5,6,7,8, 9,10	DEC All-In-1, IBM PROFS, EtherNet

Key:

B—beeper
DID—direct inward dialing
DTI—digital trunk interface
T—time/day
U—urgent

1—AT&T System 75/85 and/or Dimension
2—Centrex
3—Fujitsu Focus
4—Hitachi X
5—InteCom IBX
6—Mitel SX
7—NEC NEAX 2400
8—Northern Telecom SL-1/SL-100
9—Rolm CBX
10—Siemens Saturn
11—TIE/communications
12—Toshiba

higher level of service can be available if the system interacts with Centrex or a PBX.

Two techniques available for this are interfacing and integration. Integration works directly with PBX facilities and provides services such as:

- Activating the call waiting indicator
- Transferring callers to the PBX operator
- Forwarding unanswered calls to a personalized greeting

The use of interfaced systems makes it more difficult to leave messages. Unanswered calls usually are forwarded to the voice-messaging system, but it normally cannot determine the extension the call was forwarded from. The caller then has to reenter the extension to leave a message. This operation is only an inconvenience if the extension is the same as the mailbox address.

When the caller needs to know the mailbox address, it becomes a problem. Interfaced systems also cannot activate message-waiting indicators. The level of PBX services available depends on how open the PBX's interface specifications are. Complete interface specifications are available for Northern Telecom and Centrex systems. True PBX integration means systems that are able to activate message-waiting indicators, transfer unanswered calls to a personal greeting, and transfer callers to the PBX operator. Before installing a voice-messaging system, users should be sure that their PBX has compatible software.

The networking of voice-messaging systems is an attractive concept, but it can be hampered by several factors. Wholesale message exchange might not be possible with systems from different vendors that are incompatible. There also is a lack of network-control software for voice-messaging-system network management.

One of the many alternatives is X.400. Networking becomes more important as voice mail use becomes more extensive and mailbox holders become members of multiple distribution lists.

In a growth environment, it is difficult to keep distribution lists updated when people move or become associated with multiple nodes. Networked systems require a centralized control facility in which a system administrator can enter updates into a central database and have these changes sent throughout the network.

Most voice-mail systems use conventional analog facilities to transmit message to remote nodes. Analog transmission is slow, and the integrity of the reproduced voice is low.

Some systems can exhibit quality problems when converting digitized voice to analog form. The voice-message transmissions over analog facilities are limited to 19.2 kbits and because voice-quantization schemes need up to 64 kbit/s high volumes of messages can cause unacceptable delays in reaching the called parties.

Voice-messaging systems are more than just a convenience. They can increase productivity because people do not spend much time socializing when the recipient knows the specific reason for the call and can be more prepared when returning it. Productivity gains of 10 percent to 30 percent are possible.

In spite of some technical limitations, voice-messaging systems offer substantial benefits. As more companies evaluate the productivity benefits, voice messaging will grow and a de facto standard probably will emerge that vendors can apply until a formal standard is adopted. This could be a Consultative Committee on International Telephone and Telegraphy standard (CCITT) with widespread acceptance.

DATA-OVER-VOICE NETWORKS

Data-over-voice (DOV) systems are an alternative to full local-area networks. These systems use the in-place telephone-wire system, letting data and voice share the same transmission path. They do this by superimposing the data on top of the voice bandwidth. A private-branch exchange or Centrex system separates the inputs and routes them to their destinations.

In the past, data-over-voice products have not been popular because of low data-transmission speeds, a lack of integrated switching facilities, and an inability to handle personal-computer-to-personal-computer communications. These limitations no longer exist with such products as Teltone's TelLAN and Gandalf Data's Dovtrex.

These types of systems transmit data between nodes at 19.2 to 64 kbps over twisted pairs and at speeds of 10 Mbit/s over fiberoptic links. The input speed accepted varies from 19.2 kbps asynchronous to 64 kbps synchronous.

Some units offer a personal-computer adapter board with Network Basic I/O-System-compatible software to support NETBIOS-compatible operating systems. The software might permit personal-computer-to-personal-computer communications and support for file transfers and servers.

The personal-computer communications software can also include emulation for communications with dumb terminals. This software also can allow dumb terminal and personal-computer users to establish and maintain concurrent sessions.

A distributed architecture is employed with a star topology. It is configured by the system administrator. Menus allow the system administrator to configure each channel for such operations as transmission speed and connect/disconnect protocols. The menus also can be used to assign operation restrictions and to provide passwords and logon procedures.

The main components of DOV systems are the network interface units (NIU) and a hub. The NIU is a multiport, programmable, hardware device that interfaces the data and telephone inputs to telephone lines. The hub is usually located in a wire closet. It separates the voice and data, routing voice to the PBX or Centrex and data to the end point, which may be a computer port, file server, or terminal.

Users are typically presented with menus that assist in making connections and routing data through the network. Token-passing schemes can be used to allow a node to gain access to the network. The hub unit will broadcast the token to all NIUs, but only the one addressed can seize it. The NIU then removes the data, inserts the data packet, and passes the token back to the hub. The hub might not separate or buffer data. As each packet is passed between NIUs on the network, only the one addressed can remove the information.

Data handling is accomplished with the NIUs, which can accept inputs from up to 96 channels and statistically multiplex them over a single twisted-wire line. Different NIU versions can interface one, eight, or sixteen data-input channels.

A 96-channel NIU is typically a 19-inch wide wall-mountable enclosure that holds 6, 16-channel NIU cards and uses one wire pair to send data to the hub. The 1, 8, and 16 port NIUs can accommodate one telephone each; the 96-channel unit can be used only for data transmission.

The NIU contains firmware to accept the token, assemble the data packet, indicate the target resource addresses, and perform error correction. Each data packet that is transmitted can use appended cyclic-redundancy-check characters for the receiving NIU error-checking functions.

An NIU can operate also as a self-contained switch that directly connects devices attached to the same NIU. Intra-NIU switching is a performance advantage because it obviates the need to transmit packets to the hub, data PBX, or other switching device and back to the target terminal. Most products support RS-232 C-interfaced devices.

CONTENTION HANDLING

The following two techniques are used for handling resource contention. First, if a called resource is busy, callers are queued to it, and their positions in the queue are indicated. Second, the system administrator can assign user priority levels for access to the resource. The system administrator can even force disconnection, if required.

A call-management menu allows operators to view call (session) numbers, call status, and the names of the resources to which calls are connected. Using a multiple-session facility, users can put calls on hold, hotkey to another call, and then return to the on-hold calls without breaking communication. Some units support user hotkeying between several logical sessions.

Another capability, call interrupting, allows calls in process to be interrupted if an emergency occurs. If users are queued to a resource, they can place that call on hold and switch to another session. When the resource is available, the system interrupts the session to inform the operators of the resource's availability.

Typical data-over-voice characteristics are listed in TABLE 7-5.

Table 7-5. Typical Data-Over-Voice Characteristics.

Devices Sup ported	Up to 23,500
Maximum data rate	19.2 kbps asynchronous 64 kbps synchronous
Transmission range	Up to 18,000 feet (19.2 kbps asynchronous) Up to 3 miles (10Mbit/fiberoptic link)
Cost per port	$400 to $600

Capabilities can vary widely among the different systems. Not all systems have an interrupt capability. If users queue to a resource, they are not permitted to switch to another session and still remain queued.

Other products provide switching at the terminal interface level and furnish session management and resource-queuing services through interrupt-facility capability.

Multiplexing performed by NIUs can reduce the number of interface cards. Important also are the error-correction capabilities, number of devices, and the data rate, which allow users to construct larger networks.

8

Emerging Technologies

In this chapter we consider a number of emerging technologies that are likely to become of great importance to network designers, planners and managers.

A new information age is upon us based on a global telecommunication network. A key part of this information-based society is the communications networks. Growth has been rapid because of the demands of society and advances in technology that lead to declining costs in equipment and services

Deregulation of various industries, sophisticated networking products, and better communications software have enabled companies to conceive of applications that were not possible a few years ago.

The demand for better data, video, and other telecommunication services is causing major changes in telephone communications. The advances in microelectronic technology have resulted in low-cost solid-state memory and powerful computing devices such as 32-bit microprocessors. These technological advances have been employed in many new products for more efficient, low-cost communications. The trends in the computer and microelectronics industries have changed the telecommunications environment because of the wide use of digital communication systems. The differences between the telephone and computer and other home or office devices is becoming more and more blurred.

The telephone network will continue to be the workhorse of the telecommunications industry. The global telephone network includes over half a billion devices that handle an average of 2 billion calls a day. Telephones in the future will be much more like intelligent terminals. They will integrate features that are either not available now or are provided by separate facilities.

The Consultative Committee on International Telephony and Telegraphy will ratify the first set of Integrated-Services Digital-Network standards, and equipment companies will bring compliant products to market. The standards are designed for both voice and data needs.

INTRODUCTION

Nearly a decade in development the ISDN standards are the foundation of the next generation of interfaces to public voice and data networks around the world. Because these interfaces are based on digital electronics, they reduce both voice and data to common formats that can be combined on the link from the user to the network.

A number of factors have made cost/benefit analysis of ISDN implementation difficult and have lead to skeptical responses for ISDN. Vendor deployment and user acceptance will be slow, but ISDN is expected to have a real impact in the U.S. from 1990 on.

Effective communications strategy requires planners to evaluate ISDN. They must be able to design long-term network architectures that can take advantage of ISDN yet also be able to adopt alternatives if ISDN falters.

ISDN's roots are the telephone companies: AT&T, the Bell operating companies and the national Post, Telegraph, and Telephone administrations of Europe. These organizations were forced to develop ISDN as a response for better operational and management support for the public common carrier networks. ISDN development has been tied to the fact that digital circuits have not been available in Europe until recently.

In Europe, ISDN is being introduced as the European networks make end-to-end digital services available. U.S. telephone companies have supported digital services for some time via the digital data-service network, which is maintained separately from the basic dial-up network.

The bulk of money spent on communications goes to voice services, and early ISDN standards efforts have centered on voice, or circuit-switched, networking as the top priority.

Standards for premises equipment and interworking with private data networks were given a lower priority. The CCITT will give priority to standards for such issues as private-branch-exchange-to-private-branch-exchange interworking, packet mode standards, techniques for merging public ISDNs and private local-area networks, and network-interface standards.

In some areas such as PBX-to-PBX links, the standards will be built on protocols already developed. In other areas such as interworking with packet networks, there might be a more evolutionary path.

Over the next several years, many users will opt for integrated environments. A number of these will depend on T-1 backbones, supported by non-ISDN proprietary multiplexers and protocols. The aim will be, as always, to support the business of the corporation in the most cost-effective manner. As long as there is a plan for smooth migration between these implementations, as well as between the ISDN standard protocols and interfaces, users will not have to suffer the costs of early equipment obsolescence.

High-capacity digital channels, or T-spans, support transmission speeds of 1.54 Mbps and greater. These channels include T-1 channels, or DS1s, 1.544

Mbps and T-3 channels (or DS3s, 44.736 Mbps). Both are beginning to take an important place in networks in the U.S.

T-1s began as internal, high-capacity digital channels used since the 1960s to interconnect end offices and toll offices of the public switched networks in the U.S. and Canada. In the mid 1970s, the Bell System began selling private networks that were miniature versions of the public switched network. It was natural to offer end-users the same high-capacity T-spans for private use once digital private-branch exchanges appeared.

Until the final demise in June 1981 of Telpak (AT&T's bulk private line offering), T-spans were more of a novelty and were referred to as special or individual filings.

When T-spans first appeared as T-1s in a tariffed offering in June 1982, many questioned how the service would ever sell. It raised the requirement for a single channel used in a voice network from 3000 Hz to 64 Kbit/s, and it also required an extensive overhaul of PBXs and other equipment used in the network.

Divestiture has done much to calm these early T-1 fears, and there has been much success of T-1s among end-users. Nearly all users with large telecommunications networks will need to use T-1s.

Key among the factors to consider when comparing T-1 equipment are:
- System architectures
- Synchronization characteristics
- Bandwidth management features
- Data and voice transmission capabilities
- Network management features
- Compatibility with standards
- Hardware management capabilities
- Processor and power specifications
- Product range
- Support

Differences related to the above factors can have an impact in several ways. The cost and difficulty of managing the network varies with the degree of attention the equipment requires. Management flexibility is affected by differences in the amount and type of information the product provides to support management and engineering of the network.

The overall reliability and availability of the backbone network is also affected by product differences. T-1 multiplexer products require users to purchase additional equipment to perform functions that are standard in competing products. Compatibility with future carrier offerings such as Integrated Services Digital Networks is needed, and although manufacturers claim to be compatible with all industry standards, they might not be.

A major driving force for telecommunications technology is the existence of standards in the United States and the rest of the world. With competition and deregulation, the need for standards is greater, but they might be more difficult to attain than in the past. In the past, AT&T determined the communication standards, which were then followed by other telecommunication suppliers.

In 1974 the U.S. Justice Department began an antitrust suit against the Bell System. In January 1982, AT&T and the Justice Department reached a settlement in which the Bell System would be dismembered and the antitrust suit would be dropped. This settlement became effective on January 1, 1984 and it required AT&T to divest itself of the local telephone network in the 22 Bell operating companies. In return, AT&T would be free of restrictions to pursue other markets such as data processing.

There has been a more piecemeal deregulation of the regional Bell holding companies (RBHCs). Almost every RBHC has requested a software waiver and there has been at least one waiver in each of the following markets: office equipment, computer sales, and electronics; software; billing services; marketing and advertising; securities, insurance and real estate; paging and cellular; foreign ventures and consulting; and training technology.

The divestiture of the Bell System and the emergence of new carriers and networks has made standards even more important. In order to avoid incompatibility and fragmentation in telecommunication systems, the implementation of new international networks such as the ISDN will have to accommodate all providers of services. The breakup of the Bell System also has meant the loss of a single point of contact for planning and providing end-to-end service, potential increases in cost, more complex implementation procedures, and some uncertainty about future standards and regulation.

The integration of digital technology will be extended all the way to the actual user. The use of completely digital networks has led to the concept of the integrated-services digital network (ISDN).

Digital transmission will play an even more important role in the evolution of telecommunication networks. While switches and terminals in the past have remained analog, the new technology and declining transmission costs permits the development and use of the new types of networks and services. The use of digital switches has resulted in the integration of switching and transmission systems, commonly called an *integrated digital network* (IDN).

In addition to these technological advances, changes have also occurred because of differences in government regulation of the industry in the United States. Technological advances during the 1960s and 1970s were controlled by public policy. The innovative products were limited by the U.S. regulatory structure, and other common and specialized carriers were often at a disadvantage because of AT&T's monopolistic powers.

The deregulation of the telecommunication industry and the divestiture of the Bell System in the '80s brought far-reaching changes to the U.S. telecommunications network.

Outside the United States, most telecommunication systems are owned and operated by a government ministry or administration. Instead of the multiple public and private networks found in the United States, most countries have only a few networks. Canada and the United Kingdom allow competition in their telecommunication networks, and several countries, including Japan, West Germany, and France, are moving towards possible deregulation.

Telecommunications regulations in many countries tend to restrict the provision of domestic and international information services to private citizens and corporate users. For example, West Germany prohibits the connection of international leased telecommunications circuits to its switched dial-up telephone network. Many countries only allow the use of data modems provided by the national Post, Telegraph, and Telephone administration. A number of countries, including Japan, do not permit electronic mail service to be provided over leased lines.

The Corporation for Open Systems (COS) is dedicated to testing network products for compatibility with the International Standards Organization's Open-Systems Interconnect model, commonly known as OSI. Other groups such as the National Bureau of Standards X.400 Special Interest Group are involved in X.400 product-compatibility testing.

The increasing availability of X.400 products has brought early U.S. users of the technology new solutions to problems ranging from connecting diverse internal-mail systems to linking to outside vendors.

The X.400 standard, which addresses the interconnection of messaging networks, is incorporated into the seven-layer OSI model at the two uppermost layers, the application and presentation layers. The Corporation for Open Systems will test for X.400 compatibility, as well as for the other parts of the model, in a way similar to how packet networks products are certified as X.25 compatible.

Many of the initial members in the Corporation for Open Systems support the IBM DISOSS format with gateway or interconnections products. Connections to IBM systems can be done with an open architecture like OSI rather than proprietary SNA products. The most useful parts of SNA are LU 6.2 and DISOSS, which could be merged into OSI.

Organizations such as the European Computer Manufacturers' Association (ECMA) have indicated support for IBM's Logical Unit 6.2 (LU 6.2) as part of the OSI model. LU 6.2 is designed for program-to-program communications.

DIGITAL SPEECH AND VOICE COMPRESSION

The standard for digital speech in telephone networks is PCM at a 64 kbps rate (toll quality) PCM or pulse-code modulation, the worldwide standard technique for representing analog voice-grade signals in binary format for digital transmission. It is standardized at a 64 kbps, 8-bit-per-sample coding for telecommunications and a 16 bit-per-sampling coding for professional audio-tape recording and compact disks. In telephone networks, PCM carries uncompressed speech and does not distort data.

When transmission costs are high or bandwidth is limited as in long-haul terrestrial networks or satellite links, reduced-rate coding (compressed) for speech provides more channels, at or near toll-quality voice. The bit rate for speech can be 16 or 9.6 kbps or lower. Encrypted voice typically operates at rates of 16 kbps and less, using modems over the switched telephone network. Local loops often use low bit rates for speech because of the limited bandwidth available in existing metallic cables. The bandwidth is even more constrained if digital speech and data are provided simultaneously as in ISDN.

Any replacement for 64 kbps PCM in public switched networks must provide:

- A speech quality comparable to PCM
- The capability to maintain speech quality with multiple conversions
- The ability to handle other signals such as voice-band data

New standards will probably use 32 kbps and one of the following schemes:

- Adaptive Delta PCM (ADPCM)
- An Adaptive predictive coding (APC)
- Variable Quantum-Level Coding (VOL)

Networks will have to be designed to deliver compressed voice end to end. Only specialized switching multiplexers will be tolerable for such networks.

A configuration of point-to-point compressed circuits with PCM or analog switches between them might have sound quality worse than phone calls in many developing countries. There are few guidelines for sound quality. As shown in TABLE 8-1, a product with toll quality sound might not be tolerable in commercial quality sound implementation.

Table 8-1. Speech Quality.

Descriptive Quality	Quality	Factor	
Highest possible	10		
Good		64 kbit/s PCM	
Toll quality		32 kbit/s ADPCM	
range	8	16 kbit/s	
		Communications quality	9.6 kbit/s
Commercial		range	4.8 kbit/s
Fair	6		2.4 kbit/s
Poor	4		1.2 kbit/s
	2		300 bit/s
Lower limit of testable range			

This is because of the processor delay that the complex algorithms used require. Faster signal-processing chips will reduce the delay, but some is inherent in the algorithms.

Some toll quality schemes require 100-millisecond, end-to-end delays. Whether such a delay is intolerable is a matter of opinion. A 200-millisecond round-trip delay is still less than the 260-millisecond delay for conversations over satellite circuits that some find acceptable; others, however, do not find it acceptable.

ADPCM (adaptive delta-pulse-code modulation) is a method for compressing voice/music by at least 50 percent, usually to 32 kbps. Some implementations compress to 24 kbps. The CCITT wideband speech standard, G.722, uses a 50-Hz to 7-kHz frequency range, which takes 128 kbps under PCM and compresses it to 64 kbps.

ADPCM predicts the shape of the voice waves and transmits the difference between the measured amplitude and the expected amplitude. It is not optimal for transmitting voice-band data. The 32 kbps CCITT standard can support more than 4.8 kbps modem data, but even this has some impairment. General Electric's version is a hybrid of sub-band coding and ADPCM.

APC (Adaptive Predictive Coding) has several variations. APC-AB (with adaptive bit allocation), APC-HQ (with hybrid quantization), and APC-MLQ(with maximum likelihood quantization).

APC is a Bell Laboratories technique for compressing below 32 kbps. APC encodes the difference in amplitude between a predicated value and a sample of the speech and transmits the difference between the two. It attempts to predict more wave factors than ADPCM, which simply predicts "next amplitude". APC can reshape the resultant coding to reduce noise. The hybrids of APC stress selected factors.

Adaptive Predictive Coding—Adaptive Bit Allocation (APC-AB), was developed in the laboratories of Nippon Telegraph and Telephone Corp. (NTT). NTT uses a quadrative mirror filter to split speech signals into three sub-bands (0 to 1 kHz, 1 to 2 kHz, and 2 to 4 kHz). Then it uses APC on each band.

NTT does not manufacture products. Matsushita Electric Industrial Company built the first prototype of a codec using APC-AB and a signal-processing chip designed by NTT.

The NTT algorithm is a hybrid of APC and sub-band coding. *Sub-band coding* is a technique that splits voice into two or more frequency bands. It normally is used as a hybrid with another coding technique. In the U.S., Micom Systems is using a hybrid of APC – AB and another technique. Time-Domain Harmonic Scaling (TDHS). This technique edits human voice patterns, deleting repetitive waves in harmonics such as the "A" and "R" sounds.

Another technique is ACIT (Adaptive Sub-Band Excited Transform). This is GTE's compression technology, and it can achieve the equivalent of toll quality at 16 kbps though it is not as good as ADPCM, and can achieve good quality at 9.6 kbps.

ACIT is a hybrid of sub-band coding and transform coding that breaks the signals into blocks and encodes the transform coefficients of the amplitudes. GTE uses ACIT and time-domain harmonic scaling.

NTT has applied to the International Telegraph and Telephone Consultative Committee (CCITT) to have its 16 kbps algorithm considered as an international standard. NTT also is seeking to develop quality voice compression at 8 kbps and perhaps even lower data rates.

Another approach is APC with Maximum Likelihood of Quantization (APC-MLQ) from Japan's international carrier, Kokusai Denshin Denwa (KDD). KDD has demonstrated a hardware implementation of APC – MLQ that can deliver quality speech at 9.6 kbps and comprehensible speech down to 2.4 kbps.

Like a modem's ability to roll back from 9.6 kbps to 4.8 kbps when line quality worsens, the codec for a voice/data multiplexer should have the ability to switch to a lower compression rate if there is danger of a line blocking.

The KDD APC/MLQ approach does not use sub-band coding, but it adapts some of the principles of Linear Predictive Coding, a technique that traditionally has been used to code the shape of vowels.

Linear Predictive Coding (LPC) is one of the oldest voice compression approaches and is still used in vocoders (QV). It has been combined with waveform techniques for compression down to 300 bits/s intelligibility being the goal. The approaches require extensive data processing, including search procedures, pattern matching, and lookup tables and can require signal-processing chips with hundreds of MIPS.

TABLE 8-2 summarizes the range of voice-compression products.

CELLULAR RADIO AND VOICE COMPRESSION

Western Europe's's digital cellular radio standard includes a 16 kbit/s voice-compression algorithm. The network operates on a 900-MHz frequency with 16-kbps digitized voice signals.

The technology used to compress speech before it is broadcast is a modified version of an approach developed by Germany's Phillips Kommunikations Industrie AG (PKI).

The compression selection was made through four tests:

- Voice performance, including a subjective rating of speech quality
- Transmission delay
- The effect of multiple voices and motor traffic on a conversation
- The complexity of the equipment and implementation

The algorithm used by PKI is called *regular pulse excitation/linear predictive coding* (RPE/LPC). It was developed as a joint project between Philips and Delft University in the Netherlands. The approach uses a type of compression

Table 8-2. Typical Voice-Compression Products.

Vendor	Bandwidth (kbit/s)[1]	Introduced	Application	Quality Advertised	Technique	Formal Voice-Quality Tests
AT&T	32	1/85	M-44 service, breaks 1.544-Mbit/s circuit into 44 voice channels[2]	Toll	ADPCM	X
Aydin Monitor Systems	32	1981	T1 multiplexer, 96 voice channels[3]	Communications	Variable quantum-level companding	X
General Electric	9.6	1st Q/86	Cellular Radio/ Police/Military	Near toll	Hybrid Sub-Band coding	
Micom Systems	9.6	6/87	56/64 kbit/s multi-plexer (4 voice circuits possible in 56 kbit/s	Near toll	APC-AB and ATDHS	
Network Equipment Technologies[4]	32 or 24	1st Q/86	T1 multiplexer, 96 voice channels at 32/128 at 24	Near toll	ADPCM	32 kbit/s
Republic Telecom[5]	12 to 14	5/86	Multiplexers for 56 kbit/s circuits and for four DS0 circuits (256 kbit/s).	Equal to ADPCM	Time-domain harmonic scaling	

1. Not counting Digital Speech Interpolation (DSI)
2. Four channels lost in overhead
3. After DSI
4. Resold by IBM
5. Resold by Timeplex

known as *multipulse linear predictive coding*, which is an enhancement of LPC, one of the oldest voice-compression techniques.

LPC transmits codings for the variations in speech patterns, rather than waveforms. It was used in early vocoders and it is considered generally to lack quality. Multipulse LPCs break the speech into smaller parts than vocoders. They also use waveform encoding.

PKI's approach uses some of the techniques of *residual excited prediction* (RELP). Like ADPCM, RELP transmits a residual signal, but it is more complex and involves sub-band coding, pitch prediction, editing of selected signals, compression, and transmission.

When RELP transmits a residual signal after filtering, the filtering occurs by the type of wave that the algorithm has predicted will appear at that portion of the waveform, rather by frequency. In the PKI system, the residual signal is filtered, then every third signal is rescanned and retransmitted.

RELP is also the general technology used in IBM's approach to cellular networks. IBM's version of RELP has been reported to achieve communications quality at 16 kbps, to 7.2 kbps.

The algorithm is called *voice-excited predictive coding* (VEPC). IBM's implementation uses a medium-bit-rate coder with a 10 MIPS-equivalent bipolar signal processor.

The IBM approach uses sub-band encoding with filters. The extensive processing required can create considerable delay. One implementation requires 160-byte buffers at both input and output ends.

IBM's algorithm might go beyond its application in cellular radio. IBM obtained a patent on a voice/data packet-switched technique called Paris (Packetized Automatic Routing Integrated System).

Standardization can be a problem because only vendors using the same compression algorithms will be able to work together. Even two vendors' products using the same algorithm might not necessarily be able to work together. For example, NET's multiplexers do not work with AT&T's M-44 service, because NET gets 47 channels from a T1 and AT&T only 44.

Voice compression can let the user handle 56-kbps circuits the way T1 multiplexers handle 1.544 Mbps circuits. With voice compression, small and medium networks can have the same kind of bypass savings and control. AT&T has linked its DDS services to T1.

Voice/data connections between large and small nodes can be all digital and range from 2.4 kbps for data up to T3 rates.

A major problem in large voice/data networks is maintaining voice quality. This has been discussed earlier (see TABLES 8-1 and 8-2 for a number of different voice-compression products and techniques).

Even though voice compression can create a quality problem for data on public networks, its use can be beneficial on private networks because a private-network operator can keep the two separate and not have to put data through voice compression. Keeping data and voice apart along with the ability to compress voice allows users to manipulate the bandwidth in various ways to achieve lower prices.

One way to get toll quality without the more sophisticated algorithms is to use high-fi telephones. An IBM test of its Voice-Excited Predictive Coding (VEPC) at 7.2 kbps showed that it had achieved only a fair quality of 6.4 on a 1-to-10-point scale. With high-quality microphones, however, quality increased to 8.4.

Only recently has it become clear that the ability to compress voice telephone calls can make data circuits less expensive. Republic Telecom has run joint compression tests with AT&T and Japan's international carrier, Kokusai Denshin Denwa (KDD).

A single 56 kbps satellite circuit between New York and Tokyo was divided into four voice circuits, along with a 19.2 kbps channel for data. To carry the same information without compression would require leasing four analog lines from New York to Tokyo and four from Tokyo to New York for international communications.

A 56 kbps Intelsat IBS circuit can be leased to provide the same service with compression for about 60 percent less. The annual savings can be substantial.

Micom Systems has announced Voice/Data Multiplexer, which will allow a user to put four voice channels and a 4.8 kbps data channel into a leased 56 kbps circuit. The multiplexer can also be used for a 9.6 kbps modem circuit, a 4.8 kbps modem circuit, and three voice circuits.

Advances in fiberoptic cable system will tend to be concentrated in the following areas:

- Processes for manufacturing optical fibers with lower costs and improved properties
- Integrated optical circuits that combine both transmitter and receiver functions
- Longer wavelength source of 1500 nm or higher for lower losses
- Single-mode fibers with higher data rates and longer transmission distances
- Wavelength-division multiplexing for increased data transfer
- Power transmission over fiber cables using techniques such as infrared radiation

Transoceanic cables like the TAT-8 use single-mode fibers, four active and two backup, with automatic switching. A transmission rate of 274 Mbps is planned that will provide four times the capacity of its predecessor the TAT-7, which used copper coaxial cable. Fiberoptic submarine cable systems also are planned to link Japan and the United States.

DACS

The Digital Access and Cross-Connect System (DACS) is designed for use in digital telephone networks. It is capable of terminating up to 128 DS-1 signals and of providing digital connections to any 64 kbps channel. It eliminates the need for D/A and A/D conversion of channels and allows direct digital rerouting, dropping, and inserting using electronic switching. The switching can be initiated with a standard terminal or remotely with a data link. DACS combines multiplexing, network control, and testing in a manner that greatly reduces digital network operation and maintenance. Future plans include the wider national networking of DACS and the addition of a 2.048 Mbps capability for international gateways.

With T-1 transmission facilities more readily available, many networks with large voice/data transmission requirements can combine multiple T-1 lines to form private communications networks.

These should have the capability to switch entire T-1 composite or individual DSO channels to various locations. The switching can be done with voice/data channel banks, T-1 multiplexers with DSO cross-connect, drop-and-insert or networking facilities, or through a newer type of switch known as a private DACS.

As an option to AT&T's DACS, vendors offer switches that will permit users to establish their own private DACS networks. One of the most common T-1 channelling schemes is to divide the T-1 composite into 24 channels, each capable of carrying 64 kbps. Each of these channels is a DSO, and 24 channels make a DSI, which provides the 1.544 Mbps T-1 aggregate. To move and distribute the DSI composite, AT&T uses channel banks.

T-1 facility is offered by AT&T. The Accunet T1.5 is a point-to-point service, and customers wishing to access different locations need to purchase separate T-1 links for each target location or use a T-1 multiplexer with a drop-and-insert facility.

The drop-and-insert approach is fine for systems with few locations, but problems associated with the drop-and-insert process have restricted its use.

The AT&T service that allows Accunet T-1.5 subscribers to switch D4-formatted DS1 composites and individual DSO channels around the Accunet network is DACS. It consists of a database that holds customer data and routing information and controllers located at the AT&T DACS control center that terminate up to 128 DS1s, one of which is used by AT&T for test purposes.

The remaining 127 DS1s, which consist of 3,048 DSO channels, can be cross-connected and placed into different T-1 composites serviced by the DACS network. This allows users to target an entire DS1 or individual DSO Customer access to DACS is through AT&T's Customer-controlled Reconfiguration (CCR) facility, which is the customer database containing the end-point destinations.

CCR operates with DACs to establish the communications links. Organizations can control their own networks with products from GTE, Coastcom, Tellabs, M/A-Com, Inc., Datatel, Aydin Computer Systems, Timeplex and Wescom Telephone Products. These companies offer switching devices that perform DACS-like services. These generic, private DACS switches take the inputs from multiple D4 formatted T-1 composites from devices such as T-1 multiplexers and private-branch exchanges and switch an entire DS1 to a target location, or can extract individual DSO channels from a DS-1, build another DS1 composite, and transmit it to the designated end-point location.

The locations in a private DACS are established by users and placed in network maps that are held within the private DACS or on a personal computer.

Internal storage is used for multiple configurations, and network reconfigurations can be done through a keyboard command at a certain time of day or an event such as a T-1 link failure.

An ASCII terminal or personal computer can be used to perform the network monitoring and control. The system is user-friendly because system facilities are menu-driven, and messages showing network condition are in English.

A data bridge is used to handle subrate (below 64 kbps data transfers such as those associated with electronic funds transfer terminals. Instead of using an entire DSO channel to transfer at 2400 bits/sec between the central processor and each terminal, the bridge will combine multiple low-speed inputs into a single DSO, freeing the DSOs for additional applications.

FIBEROPTIC CABLE

AT&T began field trials with fiberoptic cables in Atlanta in 1976 and Chicago in 1977 with systems of 45 Mb/s. AT&T began using a fiberoptic link in 1983 between New York and Washington, D.C. This was the first part of the Northeast Corridor FT3C 90-Mb/s system, which was completed in 1984 and links Boston, Massachusetts, and Richmond, Virginia with a total length of 776 miles. Other intercity fiberoptic routes will use 432 Mb/s. Other carriers such as MCI are installing fiberoptic cables along railroad right-of-ways. Some of these will be single-mode fibers operating at a bit rate of 405 Mb/s, with possible expansion to 1 Gb/s. Other 400-Mb/s and 565 Mb/s systems have been installed in the United Kingdom and Japan. Future systems will tend to use single-mode fibers operating at rates of 565 Mb/s and greater.

These advances in fiberoptics can be expected to continue. In the last decade, fiberoptics has gone through three basic generations of technology:

- A first-generation made up of systems that used multimode fibers and short-wavelength sources of 850 nm
- A second generation using graded-index fibers at wavelengths of 1300 nm
- A third- generation using single-mode fibers and wavelengths of 1300 and 1500 nm

In the early days of fiberoptics, a backbone segment would have been just another point-to-point installation dedicated to one application.

Now, with the new interconnect hardware and a little planning, communications can take place over a transparent network. Almost anything can run on a fiber trunk. It depends on what type of converters are on the network. Within the next few years, most computer and terminal equipment manufacturers will offer fiberoptic I/O. Currently these converters are offered as separate systems in the form of fiberoptic modems or multiplexers.

FIBEROPTIC SYSTEM HARDWARE

There are three elements to a fiberoptic system: the transmitter, the receiver, and the cable that connects them. The transmitter is a modulated-voltage-to-light converter, while the receiver reconverts the modulated light to electrical voltages.

There will be many new, compatible add-on developments for fiberoptic local-area networks. New standards, like the 100 Mbit/s Fiber Distributed Data Interface token-ring standard are appearing. These developments will allow networks to operate at increasingly faster speeds and allow fiber to be used in other applications.

In long-haul intercity fiberoptics, every interface is T-Carrier, but building and local communications are a mixture of protocols.

AT&T and IBM support the interfaces in IBM 3270, Token-Ring, and asynchronous RS-232. Other interface applications, such as Wang CPU-to-peripheral links, can require special services to customize a network.

AT&T supports the smaller, more easily installed 62.5-micron core while IBM supports the larger, higher efficiency 100-micron core.

A network that uses $100.00 worth of 62.5-micron fiber would have to use about $250.00 worth of 100-micron fiber. It takes that much more glass to make the heavier core.

A fiberoptic backbone can be started with a section of fiber trunk that can be expanded later without additional construction costs.

High-traffic crosspoints in which the fiber backbone can be accessed can use fiberoptic patch panels as intersections. These can be in closets or equipment rooms at each building or floor. Newer connector technology allows users to design a fiber-cable network that has multiple inline cross-connects, like the wire punch-down blocks used for telephones. At some of these intersection locations, there might be some future need to branch a fiber cable into the building or floor to access new equipment.

The simplest way to expand is to employ *space-division multiplexing*. This entails no multiplexing at all, but requires putting extra fibers in the cable and using each fiber for a different parallel channel. One fiber pair can be used for IBM 3270, another for EtherNet, and another for multiple T-1 lines. There is no crosstalk between fibers, so they all act like dedicated cables.

The differences in cost between 3270 costs and fiber are shown in TABLE 8-3 based on a 64-terminal, 300-foot coax system or 64-terminal, 500-foot fiber system.

To get more signals on the same fiber cable you can use ordinary electrical multiplexing—not analog, broadband radio-frequency techniques—but digital-time-division multiplexing or remote switching (such as that used in Wang or IBM half-duplex clusters).

Another way is to use fiber as the backbone for a local net. By using digital multiplexing, a fiber pair can be upgraded to higher transmission rates.

Fiber's benefits which include dielectric properties, noise immunity, security, size, and weight, have always been good reasons to use fiberoptic cable. Its large bandwidth and distance between repeaters have become major reasons for using fiber in local networks. The demand for higher speeds is growing as the distinction blurs between public and private networks. Fiber technology was originally developed for telephone, in which it is already performance proven.

Fiber is not expected to replace twisted-pair wire in most front-end local-area networks. Twisted-pair technology is adequate to meet the front-end local network's capacity and distance requirements.

The trend of using a twisted-pair front-end supplemented with a fiberoptic back end is reflected in wiring-system offerings such as AT&T's Premise Distri-

Table 8-3. Cost-Factor Comparison for
IBM 3270 COAX vs. a Fiberoptic System.

Number of Terminals		Terminal to control Distance		
		100 ft.	300 ft.	500 ft.
32	C	2.1	5.2	8.4
	F	4.6	5.0	5.4
64	C	4.2	1.0	1.6
	F	9.2	9.6	1.0
96	C	6.4	1.58	2.52
	F	1.37	1.41	1.46
480	C	3.21	7.92	12.6
	F	6.96	7.08	7.20
960	C	6.43	15.8	25.2
	F	13.6	13.8	14.0

Key: IBM 3270 coax = C
Fiberoptic = F

bution System and the addition of unshielded, twisted-pair media to IBM's Cabling System.

Fiberoptic technology is applicable wherever its benefits are important factors in a network decision. For example, when transmission capacities are beyond 1 Mbps over more than a hundred meters, fiber should be a consideration. These types of situations are found in video applications and in high-capacity multiplexed voice channels, where high-speed data transfer is required. Such applications usually occur in back-end networks that connect multiple local network segments, hosts and processors.

FIBER STANDARDS

A standard for fiber back-end networks has been in progress by a chartered committee of the American National Standards Institute (ANSI). The group has defined a high-speed local net interface called the Fiber Distributed Data Interface (FDDI). This interface is designed to handle voice, data, and video services over optical fiber at a 100 Mbps data rate. Fiber is the only communications technology that can handle such a data rate over the 2-kilometer maximum node-to-node distance of the FDDI token-passing ring architecture.

An FDDI network can provide a multitude of service by allowing the connection of mainframe computers, mass storage, private automatic branch exchanges, bridges, gateways, and other devices to the ring. This type of service integration on one physical network is where fiber excells.

The Fiberoptic Technical Advisory Group of IEEE 802 is developing a unified fiber-cable plant that would support fiber versions of EtherNet's carrier - sense multiple access with collision detection, token bus, token-ring,and the FDDI. The objective is to allow conversion of the cable plant from one network type to another without modifications, thus retarding obsolescence.

The largest fiber user has been the telephone industry, which has been the driving force in fiber technology. The need for bandwidth and distance has resulted in the optimization of multimode fiber, first at the short wavelengths of about 850 nm and then at approximately 1300 nm wavelengths. Speeds of 45 Mbps on the earlier short wavelength systems increased to 90 Mbps and then 140 Mbps speeds at the long wavelength. The bandwidth capability of single-mode fiber passed multimode fiber during the early '80s.

In spite of single-mode fiber's higher bandwidth, multimode fiber has advantages for some users. The difference between single and multimode fiber is in the fiber core diameter, where the light actually propagates and the particular refractive index resides.

A multimode fiber core is several times larger than the wavelength of the light it carries. This allows a number of propagation paths, or modes, for light to travel through the fiber. These paths vary in length , depending on their angles of reflection. As a result, the pulses of light enter the fiber at the same time but do not emerge simultaneously at the other end. This is called *intermodal pulse dispersion*. Multimode fibers are specified according to their intermodal bandwidth.

In single-mode fiber, the core is smaller and allows only one mode to propagate. The elimination of intermodal dispersion gives single-mode fiber a higher bandwidth.

Single-mode fiber requires a high degree of skill for splicing and coupling, especially when coupling to LED sources with an emission area that is larger than the core.

Multimode fiber's large core size allows easier coupling to LEDs and simpler, low-cost connections. Multimode is usually recommended for connector-intensive applications, such as local networks.

Fiberoptic types and light sources have developed as matched sets consisting of lasers with single-mode fiber and LEDs with multimode fiber. These sets have established the capabilities of fiberoptic systems.

The important differences between lasers and LEDs are their power, spectral characteristics, reliability, and cost. Lasers have more optical power than LEDs, so they can transmit light farther without the use of repeaters. Laser power also lets users place loss-prone devices such as connectors and couplers into a fiber link. Lasers display less spectral dispersion than LED systems and can support higher system bandwidths. LEDs are generally more reliable than lasers.

In the future, lasers will be applied to single-mode fiber. This could result in a new laser-based technology for local networks with capabilities beyond those of today. Speeds of 500 Mbps or as high as 1000 Mbps could be achieved.

Current LED multimode fiber systems are usually bandwidth-limited rather than power-limited because of the distance the link can be extended. This is because of the spectral dispersion created by the LED. Because of fiber's spectral dispersion, bandwidth limitation is more severe with short-wavelength sources than with long ones. There is a marked effect at short wavelengths. This effect is so prominent that increasing the fiber's intermodal bandwidth above about 200 MHz-km will not greatly improve total system bandwidth. The 850-nm, LED-based systems are limited to about 45 Mbps over two kilometers.

In specifying fiberoptic cable, users will sometimes specify more 850-nm bandwidth. If their systems are using LEDs, however, they are getting fiber bandwidth that can never be used. Because of this, manufacturers typically specify their network cables to have low bandwidth at 850 nm, and higher bandwidth at 1300 nm, where the spectral dispersion impact is not such a severe limitation.

Because low cost and high bandwidth are the primary design objectives, 50-micron fiber could be favored over fiber of 62.5-, 85-, and 100-micron designs. Telephony development work can be expected to carry over to local networks, and it could give rise to a new generation of laser-based, local-net fiber-cabling systems. Such systems would retrofit existing LED-based systems rather than make them obsolete.

FACSIMILE

Facsimile (FAX) allows the transmission of documents using public telephones or data networks. Four classes of facsimile terminals have been standardized by the CCITT for use over the public telephone networks. These are shown in TABLE 8-4.

Group-3 terminals operate with 2.4-kbps or 4.8 kbps modems (CCITT Rec. V.27) or 7.2 kbps or 9.6-kbps modems (CCITT Rec. V.29).

The Group-4 standard is used in some advanced facsimile equipment. It is divided into three classes, according to resolution and features (see TABLE 8-5). All three classes specify automatic conversion of resolution in the receiver, the use of the International Standards Organization's seven-level Open-Systems Interconnection reference model, and the CCITT data-network interface.

The early standardization of facsimile terminals has brought more rapid growth and acceptance compared to other emerging services.

The future will see a convergence of facsimile with other forms of communications. Mixed-mode terminals could use an optical character reader (scanner) to read text on the page and facsimile to transmit graphical information.

The facsimile industry has changed rapidly. In 1978, most machines were analog units built in the U.S., with transfer rates of two to six minutes per page. Most of these were incompatible devices before the push for standards so that newer machines could be compatible.

There are seven basic steps in the digital facsimile processes:

1. A scanner checks the document and creates a representative image, called the *bit map*. (This image describes each dot on the page in a raster-scan pattern, For facsimile with a resolution of 200, dots/inch, there are 3.7 million dots on a page with 1 bit per dot.)
2. Because this number of bits takes a long time to transmit, a processor redescribes the image so that fewer bits are required, resulting in compressed representation.
3. Next, a digital-to-analog conversion is done to provide a signal that the telephone system can carry. The digital facsimile signal is converted into an analog telephone signal.)
4. The telephone call is then placed either manually or automatically, and the image signal is sent to the addressee's facsimile terminal. These steps are all performed in the transmitter part of the facsimile terminal.

Group	CC ITT Signal	Fax Speed	Compression
1	Analog	6 min	None
2	Analog	3 min	Limited
3	Digital	1 min	Complex
4	Digital	10 s	Complex

Table 8-4. Facsimile Types.

Table 8-5. Group 4 Classes.

	Class 1	Class 2	Class 3
Mandatory Resolution (bits/inch)	200	300	300
Optional Resolutions (Bits/inch)	300, 400	200, 240, 400	200, 240, 400
Characteristics	Ease of compatibility with group 3; 5-year grand-fathering of 203 × 196 bits as equivalent 200 bits/inch; modified read II compression code	Must receive from teletex; font stored; modified read II compression code	Must receive from and send to teletex; Telex font stored; must do OCR; must use mixed-mode compression code

5. At the receiving terminal the telephone signal is demodulated, or converted from analog into digital form.
6. An expander, or decompressor, restores the complete representation of the image.
7. The bit map image is printed.

About eighty percent of the dots in a typical copy are white. If the boundries of the black dots are specified, the image can be described with fewer bits. Successive rows of dots (the scan lines) are often similar, so that additional compression can be done by describing how each row differs from the last row. This is called *two-dimensional coding*. It compares each line of the scan to a reference line.

At the low end, the environment usually consists of one or two departments that send and receive images from two or three other locations on a regular basis. Specific uses include the transmission of general business correspondence. The equipment generally sends and receives fewer than 500 pages per month and costs between $1000 and $2000. These machines are compact, easy-to-use, and offer limited automation. Transmission is either 4.8 or 9.6 kbps on analog, two-wire lines. The environment is populated by facsimile machines designed for personal use, such as the Xerox 7010, the Sharp FO-200, and the Minolta MF250.

Facsimile to facsimile is the main transmission image. Products include communications boards that allow personal computers to act like facsimile machines. An example is Panasonic's FX-BIM88 facsimile board.

The low-end environment can also include digitizing scanners such as the Canon IX 12, which uses a local-area network (LAN) link for transmitting image data. Personal computers and scanners can communicate using facsimile modems or LAN cards. Applications can include the distribution of files generated by electronic publishing equipment or the remote input of image files into a centralized database.

At the high end, the environment is the focal point for multiple departments. This environment regularly sends and receives the majority of traffic in a network of at least ten locations. Multiple departments often share equipment in this environment, driving up volume and increasing the type of documents. Usually at least five departments share this environment. There is typically one well-defined use, such as sales-order entry or loan processing. These types of applications are highly automated, programmable, and are capable of multiuse image applications like plain-paper printing, document editing, or document storage. Networking using full-duplex analog circuits or digital service at speeds of 9.6 to 5.6 kbps is often needed to handle peak traffic periods.

Store-and-forward memory, auto-dialing, and broadcast might be required. Optical document storage and transmission are likely to appear in the future. Equipment used in hub environments usually transmits 1500 pages or more per month.

The different FAX environments are outlined in TABLE 8-6.

Dedicated applications for image communication involve the transmission of images between two or more locations. This application involves wide-area

Table 8-6. FAX Equipment Environments.

Number of Network Locations	Pricing Range	Typical Equipment
10 +	High	Ricoh 610 Netexpress 2100 Fujitsu 7800
5 to 10	Medium	Panaflex 600 AT Xerox 7020 Pitney Bowes 8200
2 to 4	Low	Canon 110 Datacopy Datacopy Microfax Sharp 200

communications using facsimile for intra and intercompany communications. The transmission can be digital or analog; however, analog dominates, using the public switched-telephone network.

The urgency to transmit the document is a primary consideration because it determines volume. Users with widespread locations push this application area. Most standard facsimile machines with scanner and printer in the same unit serve this area.

In multiuse applications, the user can require input scanning, printing storage, or local processing in addition to communications. The goal is to produce, distribute, or communicate image data through image capture, storage and printing.

Multiuse capability is found in high-end, plain-paper facsimile equipment. Most machines can be used as digital plain-paper copiers and can be modified with printer drivers to function as laser printers. The speed and print quality approximates that of stand-alone laser printers. These machines can cost as much as five to ten times as much as low-end machines.

Equipment with these characteristics includes plain-paper facsimile machines such as the Fujitsu 7800 and the NEC Bit IV, and microcomputer scanners such as the DataCopy.

Multiuse image communications will include more computer-based communications. Present facsimile is used for wide-area communications, but computer-based applications have both wide-area and local-area connectivity. Future growth in computer-based multiuse image communications will be driven by the ability of LANs to handle the transmission of image data.

In 1984, Federal Express started ZapMail, an electronic document-delivery service that used machines that approximated the Group-4 standard. Two years later, Federal express abandoned the service. The ZapMail terminal combined state-of-the-art features for image communications interfaces, and connection to a private X.25 data network. ZapMail was never able to generate the volume needed to make it profitable. Federal express charged a premium for the Zap-Mail service because of the quality plain-paper printing and transmission speed.

For most customers, the use of plain paper was not critical; the reliability of delivery was more important. ZapMail was faster than overnight or traditional facsimile, but was not worth the cost in the minds of the users.

The ZapMail terminals were not Group-3 compatible, so the number of sending and receiving points was limited along with the amount of traffic. Because the ZapMail network was proprietary, this limited the rate at which volume could grow.

Problems also developed in moving large amounts of image data over the X.25 network. Bottlenecks in the network slowed transmission and sometimes blocked transmission completely.

ELECTRONIC MAIL

Electronic mail is the part of any communications network that lets users send electronic messages to other subscribers on or off the network. Electronic mail has its roots in the early '70s when users of time-shared computer networks began sending messages to each other. In the past, postal services were separated from electronic communications. Even in countries where postal, telephone, and telegraph services were provided by a single government agency, these services were treated separately.

The rapid growth of voice and data-network technology resulted in the use of in-house electronic-mail services. The users of these services exchange messages through terminals that are part of a network. The messages are stored until the user logs in and requests them. In all E-mail services, users receive a mailbox, or address, and can assign passwords to their mailboxes to prevent unauthorized persons from reading their mail.

In the United States, the Postal Service developed Electronic Computer-Originated Mail (E-COM), which is a form of electronic mail in which a customer can send a text message via telephone lines to certain post offices. These receiving offices produce a hard copy that is delivered by first-class mail.

Canada has a similar electronic-mail service, called EnvoyPost, which uses the postal service and the TransCanada Telephone System. There is also MCI Mail, OnTyme, EasyLink, CompuServe, and Telemail in addition to Telecom Canada's Envoy 100 and CNCP's EOS system.

In the late '70s, computer-based messaging systems (CBMS) were not considered as serious contenders for electronic-mail systems, although they were used by a few leading-edge firms. Today, almost every Fortune-500 company and thousands of smaller ones have at least one group using CBMS. Many of the larger companies have thousands of employees with electronic-mail systems.

In electronic form it's easy for the recipient to answer, edit, annotate, or forward messages. In addition, users can broadcast messages to other users on the network. Using E-mail has several advantages. Telephone tag problems are eliminated, and it's cheaper to send large volumes of data.

For users requiring a printed copy, many E-mail services furnish copies through courier, regular mail or overnight delivery services. These services offer an advantage because users can send messages to an E-mail service center and have copies delivered to persons who do not subscribe to the E-mail system.

Public networks are open to everyone, while private E-mail network services are used only by specific groups. Of the E-mail systems listed in TABLE 8-7, only AT&T's AT&T Mail, Computer Sciences Corporations's (CSC) Infonet, Western Union's EasyLink, MCI's MCIMail, and General Electric Information Services Company's (GEISCO) Quick-Comm are private networks. Most

Table 8-7. Electronic-Mail Systems.

Vendor/Product	Telex	International	Other E-Mail	Private Networks	Message Alert
AT&T/AT&T Mail	X			X*	X
CompuServe/EasyPlex	X		X		
InfoPlex	X	X	X		
Computer Sciences Corp./Infonet	X	X		X	X
Dialcom/Dialcom	X	X			
General Electric Quick-Comm	X	X		X	
McDonnell Douglas/ ONTyme	X	X	X		
MCI/MCIMall	X	X	X	X	X
RCA/RCA Mail	X	X			
US Sprint/Telemail	X	X	X		
Western Union/ EasyLink	X	X		X	

*Unix Systems

private-systems users assign other passwords to group or departmental mail-boxes. AT&T, Western Union, and MCI do not support multiple passwords.

Most users are looking for many types of capabilities, most of which are not provided on any single system. New arrangements that provide gateways between systems have appeared that allow users to combine services like CompuServe's databases and MCI Mail's offnet delivery services.

Anyone with a personal computer can take advantage of public E-mail services. All that is needed is a Bell 103 (300 bit/s) or 212A (300 or 1200 bit/s modem) and special communications software. Besides sending conventional text messages, all E-mail services allow binary data transmission. This lets users send programs and spreadsheets.

Software is also available for interconnecting PC programs that allow the users to develop a business form using a PC program such as Lotus 1-2-3. After the form is filled out, the communications software allows it to be pasted into an electronic-mail system.

A program such as Microsoft Access can be used with a group of PCs to automate the sending of weekly sales reports to the head office for consolidation and final reporting.

Electronic-mail applications will continue to grow in the future. The industry is maturing just as the personal-computer industry did when the IBM PC was introduced. The PC industry was once dominated by products such as the Apple II, 8-bit CP/M micros, and software like VisiCalc. Many of today's electronic-mail services will undergo changes in order to meet the needs of users.

The major areas where advances are likely to occur include: 1) the integration of PC application files into CBMS, 2) line-oriented CBMSs that will be phased out as more services allow applications with PC files beyond mailbox storage and transfer and 3) the use of gateways that allow users to sign onto one system while having access to other services. The user receives a single bill as in Telecom Canada's Net. This type of service gives users access to many different database or communication services, along with the ability to devise a menu for a specific network application.

Another area destined to grow is related to gateways. This is the development of front-end communications software such as the Instant-Mail Manager by Kensington Microware for Western Union's Easylink.

Another growth area is the development of electronic data-interchange applications. This includes business-form-generation capabilities that have been line-oriented thus far in most systems.

The continued integration of PC application programs into electronic-mail will allow users to develop forms capabilities for such applications as sales management, shipping, billing, and customer services.

While network interconnection has been the principal focus of most E-mail service providers, many are now offering application packages to further increase their network's utility.

McDonnell – Douglas Electronic-Data Interchange Systems offers a variety of application packages, such as remote-order entry and sales-information management, to users of its ONTyme service.

There is a need for a single mailbox that is usable by multiple carriers. This would eliminate the need for users to check several sources for messages.

The concept of a unified messaging service might be based on Unix. It would have to support the international Message-Handling Systems (MHS) standards. AT& T uses its Message Transport Architecture, which is ASCII-based,while MHS is binary.

In the U.S., carriers compete with each other for customers, but in other countries there are almost always monopoly services. In the U.S., there is a competitive incentive to keep other carriers from access to subscribers. In other countries, because there is a monopoly, interconnections with other national systems do not reduce the competitive advantage.

Dialcom has interconnections with systems in Canada, Australia, South Korea, Puerto Rico, Singapore, Britain, Hong Kong, Denmark, The Netherlands, The Republic of Ireland, New Zealand, Japan, and West Germany. Many E-Mail services offer international delivery to Canada, Europe, South America, and the Far East.

Offnet delivery services such as General Electric Information-Service-Company's (GEISCO) Quikgram provide overnight delivery as an extension to electronic-mail service. GEISCO prints out the message on stationery with the customer's logo on the letter and envelope, and the Postal Service does the actual delivery. This service allows customers to create, store, and update mailing lists because a letter can be sent to all or some names on the list.

The service is similar to Western Union's Mailgram service. Both services are upper-case only with regular postmen delivering the message. AT&T, U.S. Sprint Communications Company, and MCI will also print messages on the user's letterhead if desired.

GEISCO also provides interfaces to in-house E-Mail systems, such as Digital Equipment Corporation's All-In-1 and IBM's PROFS and DISOSS systems. More than half of all in-house E-Mail systems use one of these implementations.

GEISCO is emphasizing applications that offer text generation, bulletin board, and other services to its Quic-Comm subscribers. One of GEISCO's offerings is its Business link package, which offers a full range of business-oriented applications, including the ability to build private databases. Another useful service is message alert. With this service, the provider calls the subscriber and tells him a message is in his mailbox if he fails to retrieve it before a specified time. CSC, AT&T, and MCI provide this service.

Portable terminals are a growing trend. Wang Information Services Corporation has licensed GEISCO's Quik-Comm service for a portable terminal with a printer and modem. The system is aimed at mobile employees such as salesmen. The terminal weighs 6 pounds and it can be used as both a remote printer and an electronic typewriter. It provides a 24-character display. The printer operates at 16 characters per second using a 24×18 dot matrix of pins. Either thermal paper or plain paper with a thermal ribbon are required. The terminal uses a 14.3 kbyte RAM and holds about nine pages of text. The modem is a 300 bps Bell 103 compatible unit. The terminal is priced at $600.

COSTS AND RATE STRUCTURES

The costs of subscribing to E-mail services are structured like those of time-sharing systems. Users pay a basic subscription fee and are billed on the number of characters transmitted or the amount of connection time. Other charges might be due according to the time of day the service is used or the data-transmission speed.

CompuServe charges for transmission speed, while CSC charges for transmission speed and time of day. RCA and The Source impose charges depending on the time of day the service is used. Lower rates apply to connections made after normal business hours.

Some E-mail providers charge extra for messages stored beyond a specified time period. Some E-mail providers offer discounts ranging from 5 percent to 30 percent based on dollar-volume usage. Vendors with a no-discount structure include CompuServe, US Sprint, GEISCO, RCA, and The Source. Unlike telephone systems, E-mail systems are not widely connected. About 900,000 E-mail mailboxes exist on public networks, and even for those mailboxes, communication is impossible unless the sender and receiver subscribe to the same network.

Interconnections are becoming more common, with providers such as MCI Mail connecting to subscribers of CompuServe's Easy Plex and InfoPlex E-mail systems. This allows users to trade messages via a direct connection of the three systems.

MCI also provides a link to France's Missive Service for International E-mail. MCI has been a leader in establishing E-mail links between MCI Mail and private E-mail systems used by major corporations and universities with its MCI Link and VAX Mail-Gate software. MCI Link handles IBM Professional Office System (PROFS) users, while VAX Mail Gate will link Digital Equipment Corp.'s All-in-1 users.

Western Union offers a transparent link between its EasyLink service and IBM PROFS and DISOSS through software that resides in EasyLink computers. AT&T also can provide links between its service and diverse E-mail systems. AT&T's link is limited to Unix-based users. GEISCO and CSC also offers links to private networks.

MCI Mail is basically an electronic-mail-delivery service, with a minor role in information services. CompuServe offers a mixture of business-oriented timesharing services, consumer-oriented electronic-informat services, and electronic messaging services.

EasyPlex is one of several ways to send and receive messages on CompuServe. Other methods include the bulletin-board systems, known as Special Interest Groups, and the CB-radio-type of chatting mode, which can be used to pass messages between two or more online users in real time. EasyPlex is a store-and-forward service that is included at no charge to all subscribers.

MCI Mail provides another way for CompuServe subscribers to get their messages out using nonelectronic options such as overnight letters or telex. CompuServe also has telex, but this must be done using InfoPlex. The sender must know the recipient's unique address because the MCI Mail directory will not be available while on CompuServe. CompuServe's subscriber directory does not offer the same convenience as the MCI directory.

In the MCI system, users need only enter the last name of the recipient at the "TO:" prompt. If there is only one subscriber with that last name, the system inserts the full name, city, and ID number in the space. If there are several possibilities, the system will list them in alphabetical order and then the sender selects the one desired.

CompuServe has been the leader in the online-information-service industry, which includes such services as The Source, Dow Jones News/Retrieval, and Viewtron. MCI ranks with AT&T as one of the dominant communications carriers, and in electronic mail MCI is in the upper half for both the number of subscribers and the volume of messages. Western Union's EasyLink processes a greater volume of messages but CompuServe has the largest amount of mailboxes, although many subscribers do not use this service. CompuServe also allows subscribers to use AT&T's Message-Access Service, which allows E-mail subscribers to call an operator on an 800 line and ask for their messages to be read to them.

Some vendors have considerably higher subscription fees than other vendors. These vendors appear to be interested only in large corporate customers, rather than the smaller user, corporate or private.

Telex subscribers can have problems when trying to pass messages from one network to another. The telex networks of TRT, Western Union, ITT World Communications, RCA Global Communications, MCI, and FTC Communications are fully interconnected, but it is not so easy to pass from one to another. Once many of the public and private E-mail carriers are interconnected, subscribers will need only one mailbox to receive messages from any system.

MCI Mail offers low prices, and though the network has been down occasionally, reliability is generally good. MCI offers the financial databases of the Dow Jones News/Retrieval service available as a menu option. CompuServe offers basic messaging and specializes in online information services. MCI Mail also offers less than EasyLink's FYI Dialcom.

The key to connectivity is the development of a common protocol, which is what the CCITT's X.400 standard will provide. A CCITT study group has been working on an E-mail standard since 1981, and a current version of X.400 is now in use in Europe.

X.400

X.400 runs as an application in the Open-Systems Interconnect (OSI) protocol and provides a link between diverse E-mail networks. It does not currently provide end-to-end service, because the process for converting documents from one E-mail format to another is left to the E-mail provider or user. Vendors can take any one of three general approaches to implementing X.400.

Some vendors implement a pure X.400 user interface that supports only services specified within the standard. Other vendors offer a modified user interface that supports X.400 plus additional services not specified by the standard. Some vendors whose E-mail products are not fully X.400 compatible choose to retain their current products and implement X.400 gateways that support less than full X.400 functionality.

Most electronic-mail systems will be interconnected using some version of the X.400 recommendation of the CCITT. Work on a directory standard is needed before it can interconnect most systems.

GTE's Telemail service in the U.S. is connected through a partial X.400 gateway to the Envoy 100 network operated by Telecom Canada. Envoy 100 and Telemail subscribers can send messages between the U.S. and Canada as an option. An X.75 gateway is provided by GTE Telenet to Canada's Datapac network. Envoy 100 appears as a host system on Datapac.

This type of interconnection requires a common directory of the two systems' subscribers and addresses. This can be done with an addressing standard that runs on the X.400 network. MCI Mail has its complete directory online, making partial addresses easy to find. EasyLink, which is available through Tymnet, has both a printed and an online directory. Western Union prints a Telex/TWX/EasyLink phone book.

LEGAL AND BUSINESS CONSIDERATIONS

Most businesses use real-time communications, such as face-to-face encounters, telephones, and telex or they use delay communications, such as couriers and postal systems. E-mail systems really do not fit into either category. They are an example of technology outpacing legal infrastructures. The rules and conventions are not yet in place to control the formation of contracts using E-mail.

For the widespread use of E-mail to occur, rules for contract formation must be established. Contract standards would enhance E-mail service acceptance. These standards will require a multidisciplinary approach bridging the areas of business, law, and technology.

Some of the issues of concern are:

- What constitutes an E-mailbox, who is responsible for its administration, and who owns its contents?
- What degree of control do E-mail users have over their mailboxes?
- Should U.S. Postal Service (USPS) rules apply to E-mail?
- Is an E-mail reply to a telex message legally appropriate as a medium of acceptance?
- Is an E-mail "receipt" contractually different from a telex answerback or a USPS return receipt?

It might seem that E-mailbox users should have the same protection as USPS P.O. box users; however, the USPS does not clearly define users' P.O.

Box rights. E-mail user agreements vary greatly between providers and contain little or no mention of user E-mailbox contractual rights and liabilities.

E-mailbox rights have only recently been considered. Consider the passage of the Electronic Communications Privacy Act of 1986 (Public Law 99-508). The 1986 Act amends the federal wiretap law, extending the privacy rights of telephone users to cover E-mail.

MCI has set a precedent when it established MCI mail E-mailboxes for Dow Jones News/Retrieval Service subscribers without subscribers' prior consent. MCI subsequently introduced an annual E-mailbox fee, terminated the E-mailboxes of delinquent Dow Jones subscribers, and deleted the contents of their E-mailboxes without prior notice. If E-mail users have common-law rights to the contents of their E-mailboxes, then MCI Mail's actions might have violated them. A comparable action by the USPS on the contents of a P.O. box such as destroying first-class mail would constitute a felony.

In the past, telecommunications media were discrete and mutually exclusive. Telephone, telex, and other data communications were not integrated.

New technologies physically connect previously discrete entities. Protocol converters, gateways, and other devices permit dissimilar networks to interconnect. Often the user is unaware of these interconnections and their consequences. Consider the following:

- E-mail-to-facsimile communications typically link both delayed and instantaneous media, and these interconnections have text-to-graphics conversion limitations.
- Text-to-speech systems can make it difficult for users to distinguish between real voices and stored speech.
- Expert-based programs can filter and delete mail from an E-mailbox and automatically reply to specified E-mail messages.

Legal rules are needed to determine when an offer and acceptance are effectively communicated in order to form a valid contract. If the parties use an instantaneous communications medium, an offer is deemed communicated when spoken. If the communications medium is delayed, an offer is deemed communicated when received rather than when dispatched, unless carried by the USPS. The Mailbox Rule is based on common law, which says that an offer or an acceptance using USPS is effectively communicated when physically placed in a USPS mailbox. This rule is difficult to apply to hybrid E-mail transactions.

Many E-mail systems permit users to scan or preview their E-mail before reading the body of each message. Typically the scan will list the sender and the subject or first line of the message. Thus, without generating a return receipt, an E-mail recipient possibly can determine a message's content.

There is also an absence of common terminology for defining E-mail transactions. Vendors use the following terms to indicate the initiation of E-mail: item created, posted, sent, stored, accepted, queued, and transmitted.

Some systems automatically reply not only to the sender, but to everyone on the original message's distribution list. Other systems generate a reply only to the sender.

Once rules and standards for E-mail are developed, its uniformity will reduce these contractual uncertainties and enhance its acceptance for important communications that are usually communicated via overnight courier or other more traditional methods.

The Accredited Standards Committee X.12, operating under ANSI, is developing standard Electronic-Data-Inerchange formats that will let users electronically communicate business forms such as freight bills, invoices, and purchase orders.

The Consultative Committee on International Telephony and Telegraphy has developed standards that indirectly affect contract formation rules. These include standards for message handling (X.400), Integrated-Services Digital Networks, and a draft standard for a global, internetworked, electronic-directory service. This directory would identify each user's communications facilities, hours of operation, speed, supported protocols, delay associated with its data-termination equipment, and business and personal information. The Electronic Mail Association, which was instrumental in introducing and passing the 1986 Communications Privacy Act, is also addressing E-mail issues.

Presently, E-mail is relegated to supplementing intracompany communications. As it grows in user base and types of applications, E-mail will be held accountable to a higher standard and should become usable for most types of communications.

VIDEO TELECONFERENCING

Video teleconferencing allows users at two or more locations to have real-time, interactive, visual contact. The images that are transmitted could also consist of presentation graphics in addition to the speakers. If the movement can be limited, bandwidth compression techinques could be applied while still achieving sufficient resolution. Most video teleconferencing services use a combination of digital encoding, bit-rate reduction, and digital transmission to lower transmission costs. Bit rates could range from 1.5 to 6.3 Mb/s.

Standards for video teleconferencing have been based on video broadcasting standards, such as those established by the U.S. National Television Systems Committee. The trend in video teleconferencing has been towards lower bit rates to lower costs, a single international standard for compatibility, and the integration of video with voice and data signals.

AT&T's Picturephone Meeting Service (PMS) has used rates from 1.5 to 3 Mb/s. PMS is a part of the high-speed, switched digital service (HSSDS) offered by AT&T between major cities. Similar video teleconferencing services are offered by other carriers, using satellite and terrestrial networks.

Satellite Business Systems (SBS) offers Video Network Services (VNS), which is a teleconferencing service that provides full-motion, television-quality broadcast video. Satellite systems are used in a point-to-multipoint configuration.

An earlier service, SBC Videolink, digitized the video signal and used data-compression techniques, but it did not provide full-motion video. The VNS service transmits an analog video signal. Videolink was point-to-point only and used company-owned meeting rooms in major U.S. cities.

Services like VNS allow large companies to disseminate information quickly to geographically dispersed locations. They use small-aperture earth stations in the range of 1.8 to 2.8 meters. Interactive communications might be limited in some services. VNS, for example, can return only audio from a remote site to the original point of broadcast, which is done using terrestrial and not satellite communication. Services that offer two-way data communications via satellite are available from Video Satellite (VSAT) providers.

VIDEO SATELLITE SERVICES

VSAT technology is envisioned as a replacement for leased-line networks. Users get end-to-end digital service. The cost of a long-distance carrier point-of-presence-to-customer-premises segment of a long-distance link has risen sharply.

VSATs can be installed faster than leased lines. Predicting the installation of leased lines is an imprecise art and can be difficult. A VSAT can be installed anywhere in the U.S. within about 30 days.

Interactive-satellite data-communications networks are based on techniques that combine both data processing and telecommunications technologies. While 19.2 kilobits-per-second is still being used as the data transmission speed for earth stations, most newer systems will use T1 technology at speeds up to 1.544 megabits-per-second.

Users have the option of purchasing a private VSAT network and operating and maintaining the system themselves or of subscribing to a shared VSAT service (see TABLE 8-8). The majority of VSAT systems are shared with multiple users using a single hub that serves as the center of the VSAT net. Most users have opted away from private networks because the capital investment is high.

Cable TV (CATV) and direct broadcast by satellite (DBS) are common in the United States and Europe. The transmission of video signals for TV is expected to remain analog for the near future, but two-way cable TV based on digital transmission is already in use. Subscribers are provided with a low-speed digital connection for these interactive services in addition to the normal one-way video channels.

Table 8-8. Two-Way VSAT Services.

Vendor Service	Shared or Private Hub	VSAT Size (Meters)	Data Protocols Supported	Ku- or C-Band Usage
(Networks of more than 100 VSATs)				
American Astronet Resells CTP	Shared	1.8	SNA/SDLC, X.25, bisynchronous	Ku
AT&T Skynet Star Network	Both	1.8	SNA/SDLC, X.25, BSC, asynchronous, DEC DCMP, Unisys, poll/select	Ku
COMSAT technology Products (CTP) Starcom Data Service	Shared	1.2, 1.8	SNA/SDLC, X.25	Ku
Starcom Network Service	Private			
Contel ASC Small-Station Service	Both	1.8	SNA/SDLC, X.25, BSC 3270, asynchronous, Unisys poll/ select	
Equatorial Communication Equastar Transaction Network Service	Both	1.2	SNA/SDLC, X.25, BSC, Asynchronous, Unisys poll/select	C
100 VSATs)				
GTE Skystar Inter-active Network McLean, VA	Both	1.8	SNA/SDLC, X.25, asynchronous, bisynchronous	Ku
Networks of Fewer than 100 VSAT,				
Nova-Net Communication Resells Equatorial Service	Shared	1.2	SDLC, X.25	C
Skyswitch Satellite Communications	Uses VSAT-to-VSAT design, no hub needed	1.8	X.25	C, Ku

Key: BSC —Binary Synchronous Communications   SDLC—synchronous data-link control
HDLS—high-level data-link control   SNA—Systems Network Architecture

Although several companies offer a single point of contact for all parts of the service, few actually own the transponder space, manufacture the VSATs, and operate national field-service organizations. K-Mart's VSAT network is an example. GTE Spacenet served as the point of contact and transponder space supplier. NEC supplied the small earth stations and GTE Spacenet contracted with a third-party firm to provide maintenance for the VSATs. Telenet provided the network communications center.

AT&T, CTP, and RCA conduct operations in the same manner. Equatorial and Contel ASC offer single-vendor solutions. Users should be near their VSAT service vendor's shared hub station because they might have to run at least one leased line from their data center to this hub (FIG. 8-1).

VSAT users who do not own their buildings must gain landlord approval for a VSAT on the roof or side of the structure. Local zoning restrictions can also apply.

Users also need a clear line of sight between the VSAT and the satellite at each location. This applies only to the microwave and infrared systems, but it is an important issue that needs to be considered when installing a VST net.

Maintenance and continuing field support are also important when selecting a VSAT network. Users have a choice of purchasing equipment and service from vendors that maintain national field forces or from vendors that subcontract the work to a third party. Each approach has its merits. A VSAT vendor's own support group might be more familiar with the vendor's product, but a third-party group might possess a more diversified knowledge of VSAT maintenance.

Users can test this wide-area technology with a small network that typically runs form 30 to 90 days. This pilot test enables users to evaluate small-earth-station technology for specific applications.

Some users opt for full-blown VSAT networks. Prudential – Bache Securities plans to create a 300-site network to handle data, teleconferencing, and broadcast video applications from its New York headquarters to its branch office sites. S&L Data, a provider of communications services to financial institutions,

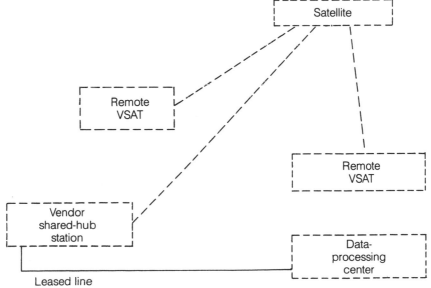

Fig. 8-1. Share hub VSAT.

chose Equatorial for a 15-state VSAT network to serve most of these institutions.

Equatorial and Telenet Communications are providing the U.S. Department of Agriculture with a 280-site VSAT network for its Forest Service. In all of these cases, replacement of costly leased lines was a key reason for selecting satellite networks.

DIGITAL MICROWAVE

Digital microwave offers a wide range of transmission facilities (TABLE 8-9). Users can select from units providing T-1 (1.544 Mbit/s), T-2 (6.312 Mbit/s), and T-3 (44.73Mbit/s), broken down into multiple T-1 circuits and DS-X data-carrying capacity. The transmission technique uses time-division multiplexing to segment the bandwidth and employs the familiar T-1 DSO 64 kbit/s channelization to parcel out time slots to each input. A DS1 channelization scheme, used with T-1, has 24 DSOs, while a DS2, or T-2, consists of 96 DSOs. T-3 uses a DS3 and consists of seven DS2s.

The major reasons to bypass the local telephone company are need and economics. If user needs change to the point where current carrier facilities are inadequate or too expensive, bypass can be the solution.

T-1 transmission facilities have been a tariffed offering since 1983, but some customers still have to wait six to twelve months for installation if they can get it at all. Not every local operating company offers T-1.

Digital microwave radios are a popular bypass alternative for those who need high-bandwidth voice and data. A company can operate a digital microwave radio if it has an FCC license. Users must perform a frequency analysis to ensure their operating frequency does not interfere with anyone else. A frequency analysis requires detailed knowledge of current frequencies in use and those proposed.

Before installing a digital microwave radio, the user must conduct a site survey to determine if anything will interfere with the transmission. The obvious problems are obstructions like buildings, trees, and hills. Less obvious are mirrored buildings, rivers, and lakes, which can cause wave-reflection problems.

The site survey also considers atmospheric conditions such as the frequency and intensity of rain, fog, snow, and humidity, because these can cause excessive transmission errors. The lower the transmission frequency, the less effect atmospherics will have; higher frequencies like 23 GHz can have problems with heavy rain.

Most vendors use computer programs to calculate the effects of factors such as transmission length, power, weather, free-space transmission loss on signal fading, and the resulting errors.

The complexity of site surveys depends on the distances involved. For 23 GHz, it is mainly line-of-sight evaluations because the distances are short.

Table 8-9. Digital Microwave Systems.

Vendor/Product	Frequency	Transmission Capacity (Bit/sec)	Maximum Number of Voice Channels	Modulation Technique	Bandwidth	Bits Per Hertz	System Gain
AT&T/DR-18	17.7 to 18.14 GHz 19.26 to 19.7 GHz	44.73 M	672	MSK	40 MHz	1.12	100 dB at 10^{-6} BER
DR-23	21.2 to 23.6 GHz	1.544N or 6.312 M	192	AM	10 MHz	0.65	97 dB at 10^{-6} BER (for f−1); 90 dB at 10^{-6} BER (for T-2)
Avantek/ DR2D-192	1.7 to 2.3 GHz	12.75 M	192	QPR-7	3.5 MHz	3.7	110 dB at 10^{-6} BER
DR23 A	21.2 to 23.6 GHz	1.544 M	96	FSK	25 MHz	NA	96 dB at 10^{-6} BER
Aydin Microwave AD-4	4.4 to 5.0 GHz	Up to 18.72 M	300	QPSK	10 MHz	2	114 dB at 10^{-6} BER
Digital Microwave Corp./ DMC-23	23 GHz	1.544 M or 6.312 M	96	N/A	21.29 to 23 GHz	0.25	97 dB at 10^{-6} BER (for S1); 90 dB at 10^{-6} BER (for DS2)
Harris/DVM 2-45	1.85 to 1.99 GHz	45M	1,344	64 QAM	10 MHz	4.5	104 dB at 10^{-6} BER

Table 8-9. Continued.

Vendor/ Product	Frequency	Transmission Capacity (Bit/sec)	Maximum Number of Voice Channels	Modulation Technique	Bandwidth	Bits Per Hertz	System Gain
Urbanet 10	10.5 GHz	6.312 M	96	QPR	2.5 MHz	2.6	88 dB at 10^{-6} BER
Loral Terracom TCM 620	1.74 to 19.79 GHz	1.544 to 44.736 M	24 to 48	4 PAM or 3 LPR	1.7 to 19.7 GHz	1.6 with 4 PAM; 1.1 with 3 LPR	NA
Motorola/ Starpoint D-2000	2 GHz	12.35 M	192	7 LPR	10 MHz	1.3	106 dB at 10^{-6} BER
NBC/DMR-69 – 119 – 135 MB	6 to 11 GHz	135 M	2016	64 QAM	30 MHz at 6 GHz; 40 MHz at 11 GHz	4.5	102 dB at 10^{-6} BER f0r 6 GHz; 97.7 dB at 10^{-6} BER for 11 GHz
Northern Telecom/ RD-6	6 GHz	135 M	2016	64 QAM	23 MHz	4.5	101 dB at 10^{-6} BER
Racon/Micro-pass 7015	21.2 to 23.6 GHz	6,312 M	96	NA	33.8 MHz	2	172 dB at 10^{-6} BER
Rockwell Intl./MDR-2306	5.9 to 6.4 GHz	135 M	2016	64 QAM	30 MHz	4.52	101 dB at 10^{-6} BER; 99 dB at 10^{-8} BER

KEY:
AM —amplitude modulation
BER—bit error rate
FSK—frequency-shift keying
LPR—level partial response

MSK—minimal shift keying
NA —not available
PAM—pulse amplitude modulation

QAM—quadrature amplitude modulation
QPR—quadrature partial response
QPSK—quaternary phase-shift keying

Longer distances are more difficult because all obstacles might not be that apparent.

The growth of digital radio also is connected to the rapid spread of digital cable systems. Both the T-carrier and fiberoptics types can be interconnected easily via digital radio.

The digital radio improvements have produced spectral efficiency, multipath fading at high transmission rates, and reliability. A spectral efficiency of 4.5 bps/Hz has been achieved with 64-QAM. Other coding and modulation techniques such as the Ungerboeck codes could produce greater bandwidth efficiency. New types of adaptive equalizers, such as those that use transversal equalization, allow transmissions rates in excess of 100 Mb/s. Reliability has been improved with the replacement of TWTs with solid-state devices.

Most users let a common carrier handle long-distance transmission needs. They either tie their microwave into a local telephone company and let it feed the data to the common carrier, or they bypass the local telephone company and directly interface with the carrier's point of presence (POP). If the latter is chosen, it can require the customer to mount an antenna on the carrier's roof and place multiplexing equipment at the site.

Current microwave units are solid state, and component failures are infrequent. All vendors offer maintenance contracts that are less expensive than the tariffs users are used to paying.

T-1 SERVICES

T-1 is a means of physically transporting a 1.544 Mbit/sec digital signal between two points. This is confused sometimes with a DS1 signal, which is a 1.544 Mbit/s digital signal that is carried over T-1 carrier systems. T-1 refers only to the bit rate and not to the signal content. DS1 has a specific format. T-1 multiplexers satisfy the bit-rate conditions and utilize a bipolar alternate-mark inversion (AMI) signal.

The most common use of local T-1 channels is for access to a long-distance carrier's point of presence (FIG. 8-2). These local T-1s can be acquired from the local operating company or bypass vendors.

On- site multiplexing equipment or the capability in a PBX can be used to combine the following onto a number of T-1s:

- Long-haul voice-grade channels used in private voice networks
- Foreign-exchange lines
- Lines used to access off-net services such as WATS

At the carrier's point of presence, the T-1s, are broken into individual voice-grade channels for routing. They can also be packed into other T-1s for intercity transmission.

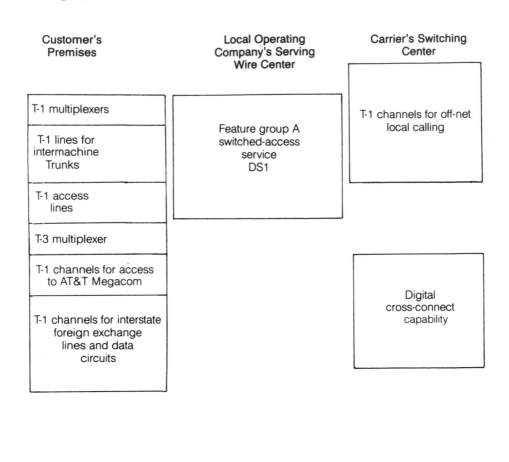

Fig. 8-2. Typical use of T-1 local channels.

At the carrier's switching center, the packing and unpacking process is known as *M-24 multiplexing*. Here 24 voice-grade channels are packed onto a single T-1. A physical device such as a D4 channel bank might do the multiplexing and demultiplexing or software techniques could be used. Digital cross-connect services simplify the process of unpacking the incoming T-1s and rerouting individual channels to be packed onto outgoing T-1s.

Other T-1 multiplexing techniques use statistical and error-correcting methods to deliver to toll-quality voice at bandwidths of 32 kbit/s and 16 kbit/s.

At 32 kbit/s, a single T-1 can carry 48 voice-grade channels; at 16 kbit/s, it can carry 96. One 32 kbit/s technique is M-44, which combines two T-1s onto a single T-1 channel. This is also called *channel-expansion multiplexing*.

Because of the encoding scheme used in M-44, four subchannels cannot be used, and the resulting T-1 can carry only 44 voice-grade channels. Other T-1 multiplexing techniques allow better utilization.

Analog voice-grade data circuits also can be multiplexed onto T-1s. The easiest way is to put each circuit on one of the 24 subchannels of the T-1. This approach is known as *voice-band data*. The entire 64 kbit/s subchannel is used to carry a data channel that might be operating at only 2400 bit/s.

A better approach is used to pack the 64 kbit/s subchannel. The analog data circuits are converted first to digital form, then multiplexed with other data circuits onto the 64 kbit/s segment.

It is possible to merge data and voice on the same T-1. However, data circuits are typically multipoint polled lines that can be tuned to maximize performance for a particular communications protocol. The control and synchronization data that are routed with the data must be sent through devices that are tuned for voice communications. There is a risk that this control information could be degraded, causing errors in the data network.

T-3 SERVICES

T-1s are not the only T-spans available. There are also T-1C (3.152 Mbit/s), T-2 (6.312 Mbit/s), T-3 (44.736 Mbit/s, and T-4 (274.176 Mbit/s) services. Fiber bundles also have been installed for customers at speeds in excess of T-4. AT&T introduced T-3 service (Accunet T45) over its fiber network in 1986.

T-3s have the capacity of 28 T-1s and can replace five to fifteen individually purchased T-1s. In large networks, T-3s can be cost-effective in carrying local channels to a carrier's point of presence.

BANDWIDTH ALLOCATION

The early T-1 multiplexer networks assigned bandwidth at installation and never changed it. The only way to stop a channel from using bandwidth was to disconnect the channel card and remove it from the multiplexer's timing frame. This reframing process requires manual intervention.

Improved multiplexer and nodal processor equipment offers a more sophisticated means of bandwidth conservation and management. By dynamically allocating bandwidth, communications managers capacity can be minimized between system nodes. Some techniques involve only voice, some only data channels, while others involve the complete network.

Older T-1 multiplexers controlled capacity to any channel that was connected, whether or not it was needed, using simple time-division multiplexing. Improvements can be made if data channels are given capacity only when the attached processor equipment or the user requests it with the data terminal ready (DTR) connection.

The T-1 multiplexer or nodal processor can also monitor the voice channels, committing network capacity only when the voice systems connect a call over the network. The trunks that connect voice private-branch-exchange systems usually are sized to be completely full for only one out of every 100 calls during the busiest part of the day. During the rest of the day, many of the network voice circuits are not in use, and their capacity can be used to support other services.

Rather than continuously requiring the maximum design capacity for each channel, some of the newer nodal processors use a virtual-circuit approach.

DOWNTIME CONSIDERATIONS

Because a T-1 network places so many communications channels on one physical circuit, circuit failures have a greater impact on operations than similar failures in analog circuits. Several techniques exist to minimize the impact of a T-1 circuit failure. One of the most common is spreading critical traffic over as many T-1 circuits as possible. This allows the least disruption to critical applications when a T-1 circuit fails.

T-1 processors have the ability to perform automatic channel rerouting on a priority basis during T-1 circuit failures. Among the techniques to minimize circuit failure impact are providing multiple parallel routs between critical T-1 nodes and grouping traffic into routing categories, such as:

- Critical traffic that requires on-network backup
- Traffic that can be switched to an alternate network
- Preemptible traffic that can be disrupted

Noncritical traffic can be disconnected from the T-1 network without significant impact.

Most networks contain a significant amount of noncritical traffic. By carefully balancing the mix of each of these categories on all T-1 circuits there can be adequate bandwidth on the surviving T-1 circuits to accommodate all critical traffic from the failed T-1 circuits.

The key to providing cost-effective networking is to fill all T-1s as completely as possible, but this cannot be done if users claim that all of their traffic is critical. An effective way to solve this problem is to charge users a higher rate to provide on-net backup and allow the users to decide if they want to pay for its benefits.

ISDN

The changes in the switched-telephone-network marketplace and the growth in digital communications have resulted in the emergence of the concept known as the integrated-services digital network (ISDN). The concept is based on the use of a limited set of user interfaces and protocols to provide end-to-end digital connectivity for a wide range of feature-rich integrated voice and nonvoice services.

The CCITT has been active in defining and standardizing the ISDN concept. The complete transition from existing networks to a comprehensive ISDN is expected to take place over the next 5 to 15 years.

ISDN can use the present integrated digital networks (IDN) that employ 64 kb/s switched systems. Services can include voice, data, facsimile, teletext, and videotext.

As shown in FIG. 8-3 initially, the ISDN concept will use two basic types of communication channels:

1. A B or bearer channel at 64 kb/s for the transmission of digital voice and data.
2. A D channel at 16 kb/s for out-of-band signaling to support the B-channel services. The premises controller at the customer site will be a PBX or a modified T-1 multiplexer.

Without video services, two B channels and a D channel are needed for a total of 144 kb/s. Adding more B channels allows higher speeds and moves the D channel up to 64 kb/s.

EVOLUTIONARY TECHNOLOGY

The ISDN of the '80s, featuring 64 kbit/sec basic rate access, will not be the end of network technology development. Under development are the 100 Mbit/sec Fiber Distributed Data Interface, 135 Mbit/sec-plus wideband ISDN, and high-speed, hybrid circuit/packet switches.

Users should plan for the ability to interwork common-carrier services with ISDN and evaluate the role ISDN can play. In some applications, ISDN PBXs connecting integrated voice/data terminals might be preferable to separate PBX and local-network wiring schemes, at least at relatively low data rates. In other areas, high-capacity, dedicated, local-area networks could be preferred. In

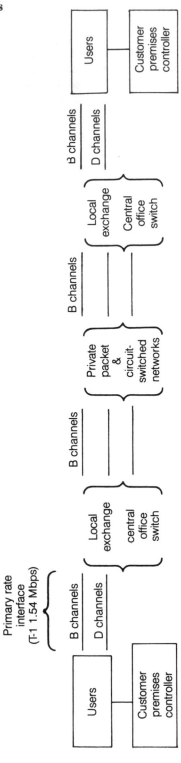

Fig. 8-3. ISDN concept.

either case, users should buy from vendors that have plans for accommodating ISDN in the 1990s.

The next several years will be a period for considering ISDN's impact on corporate networks. Equipment should be upgradable.

ISDN is a technology that will grow with user acceptance. Customers will be responsible largely for building applications for ISDN technology, because they best understand their own needs.

The implications of ISDN are great. The technology can make possible many things that were previously impractical or impossible and can pave the way for many networking applications. Companies that recognize the potential of ISDN early might be able to gain business advantages over their competitors.

While technologically possible, it is not economically feasible to build a single integrated network given the billions of dollars worth of existing telephone network equipment already in place. The ISDN concept does not presume that this country will see the development of one monolithic, integrated, digital-transport and switching system. The fragmented and competitive nature of the communications industry has resulted in varying degrees of commitment to ISDN

CIRCUIT AND PACKET SWITCHING WITH ISDN

Users making premises equipment decisions for ISDN need to review the past deployment of new technology. The question of basing premise communications architecture on existing technology such as LANS or on ISDN technology involves both circuit switching and packet switching.

Circuit switching is the technology used for the public telephone network and has been in use for over 50 years. Packet switching was introduced in the late '60s as an alternative to dedicated or dial-up circuits.

Circuit-switched service provided the ability to connect to different destination end-points, one at a time, through the same access line. The telephone is not hardwired to a single destination but can be used to call any endpoint on the public switched-telephone network. You use your telephone to carry on only one conversation at a time, except when conference calls are set up.

Packet switching was the first major alternative to dedicated circuits and dial-up service as methods of transmitting data. Packet switching has the ability to connect to multiple locations simultaneously using the same line.

ISDN relies upon digital transmission using out-of-band signaling technology, which provides greatly increased intelligence over the public telephone system. Trials of ISDN access technology have been scheduled by all of the regional holding companies. The first phase of commercially available ISDN services will focus on improved circuit-switched services.

Many trials concern the basic rate interface in a Centrex environment. The basic rate is 144 kbits/s divided into two 64 kbits/s information carrying B channels and one 16-kbit/s signaling and information-carrying D channel. Benefits include:

- Voice/data integration on a single access line
- Coaxial cable elimination since circuit-switched data in ISDN uses existing Centrex wiring
- 64 kbit/s facsimile

Some trials also include the primary rate interfaces 1.544 kbit/s divided among 234 64-kbit/s B channels and one 64 kbit/s D channel.

The goal of ISDN is to integrate the access to circuit and packet-switched services, permitting a single user terminal and wiring system (FIG. 8-4). It provides important advances over basic circuit switching. It allows more than one circuit-switched connection over a single user interface, similar in some ways to packet switching.

A computer can connect to an ISDN switch using the primary rate interface for 23 separate 64 kbit/s circuit-switched exchanges at one time.

ISDN will allow protocol-based control to be extended to circuit-switched services through the D channel.

ISDN will not change the underlying transmission networks. Packet and circuit-switched networks will continue to function independently. The technology to packetize all communications, including voice, depends on the high throughput and low error rates associated with the fiberoptics and might be deployed commercially when narrowband ISDN and fiber are available in the late 1990s.

PBX circuit switches provide a connection service for data equivalent to dial-up. PBXs provide some services that are designed for data transmission. Manufacturers have developed nonblocking PBXs that have the capacity to enable all endpoints to use the switch simultaneously. They will still dedicate capacity rather than share resources as in a packet environment.

LANs are protocol-based and provide error and flow control in addition to sharing the communications medium. Like packet networks, they allow for multiple, simultaneous connections and do not block.

STANDARDS AND IMPLEMENTATIONS

Current standards call for true packet-switched connectivity as packet-handler modules are added to ISDN central systems. The connection of ISDN-based equipment and local networks can be more difficult. The 1.5 Mbit/sec-to-2 Mbit/sec range primary rate of ISDN is exceeded by the throughput supported by local networks. This throughput is currently 10 Mbit/sec and will soon reach 100 Mbit/sec as the use of fiber-based local-network technology grows.

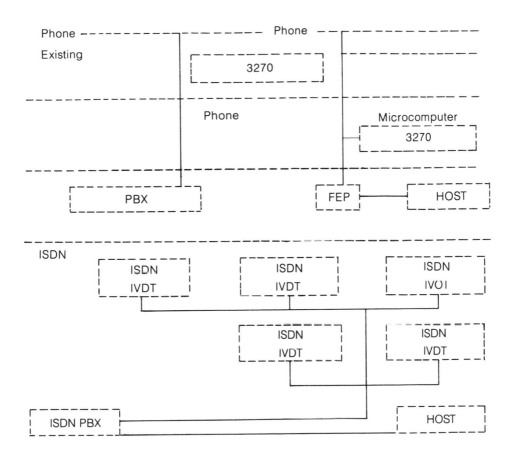

FEP — front-end processor
ISDN — integrated-service digital network
IVDT — integrated-voice/data terminal
PBX — private-branch exchange

Fig. 8-4. Premises architecture.

Standards for local-area networks and ISDN interworking will probably appear when wideband ISDN channels are defined.

AT&T, which is farther ahead in its ISDN effort than any other carrier, is implementing its integrated digital network as a collection of independent special-purpose nets terminated at AT&T switching nodes. Within this nodal ISDN architecture, the only thing actually integrated is the link between customer and network. ISDN addresses the integration of access. These multichannel integrated access links will do away with the need for separate voice and data trunks and special trunks for accessing specific services.

The channels within the access links will be controlled by ISDN-compatible customer- premises equipment working with carrier-based switching systems.

Vendors such as AT&T and Northern Telecom have suggested extensive integration schemes among home shoppers, store-based telemarketing retail-chain buyers, manufacturers, and analysts setting prices. ISDN could provide these parties with instant access to information via existing telephone networks. Current ISDN trails are on a simple replication of existing services.

ISDN will be adopted slowly, in an evolutionary manner. For some users, ISDN might eventually provide a flexible premises architecture. For others, dedicated LANs might be more important.

ISDN might not succeed as a data technology any more than did PBXs. LANs have disadvantages in that they are not suitable for voice, can require special cabling, are experiencing changes in standards, and will continue to evolve rapidly, forcing periodic equipment upgrades and/or replacements.

Major users are considering limited investment in ISDN experimentation. Participation in ISDN trials, particularly those involving a wide range of premises equipment, is a way for users to determine the applicability of ISDN to their own network architectures and to have an influence on the further development of standards.

Several trial participants have eliminated coaxial cable by using ISDN's twisted-pair cabling to link 3270-type terminals and personal computers to mainframes. (Some initial ISDN implementations are shown in TABLE 8-10).

Table 8-10. Initial Domestic ISDN Implementation.

Telephone Operating Company	Location	Customer	Date	Switch	Access Method
Southern Bell Telephone & Telegraph Co. BellSouth Corp.)	Atlanta	Bank of Georgia	3/88	AT&T 5ESS	2B + D, 23B + D
	Boca Raton, Fla.	Prime Computer	7/88	Siemens EWSD	2B + D, 23B + D
Southwestern Bell Telephone (Southwestern Bell Corp.)	Houston	Shell Oil Tenneco	Mid-88	AT&T 5ESS	28 + D
	St. Louis	AT&T	3rd quarter 1988	AT&T 5ESS	2B + D

A prime target initially will be implementing ISDN in a sales office. These applications are data-oriented and include casual and heavy data users accessing host computers via terminals and personal computers.

Tenneco plans a more wide-ranging operation that includes the oil company's headquarters as well as the headquarters for its various divisions. Initial ISDN application offerings are shown in TABLE 8-11 with those implemented indicated by an asterick (*).

Two standard ISDN access-link configurations have been defined by the Consultative Committee on International Telephony and Telegraphy: a primary-rate interface and a basic-rate interface.

The primary-rate interface is intended for business environments that support high-capacity devices such as private-branch exchanges. It specifies a single T-1 1.54 Mbit/s digital communications link divided into 24 individual 64 kbit/s channels. Of these 23 are B, or bearer, channels that can carry voice or data. The twenty-forth is a D or data channel that carries messages controlling the B channels.

The basic-rate interface is intended to support terminal and other equipment used by small businesses and residential users, as well as end-user devices in Centrex environments.

The basic rate specifies a single 144 kbit/s facility segmented to provide two 64 kbit/s B channels and a 16 kbit/s D channel to provide out-of-band signaling and a separate signaling channel that carries the information required to establish and monitor voice and data calls over the B channels.

Besides signal messages, the D channels of both the primary and basic-rate interfaces will accommodate packet-data communications. Part of the D channel can be used for packet-switched data, or a full B channel can carry packet data.

AT&T and other long-haul carriers will initially support primary-rate interfaces, while the Bell operating companies and other local-exchange carriers are focusing on the basic-rate interface.

Table 8-11. Initial ISDN Applications.

*File Transfer
*Local/Wide Area Networking
*Packet Switching
Coaxial Elimination
*Line Consolidation
Modem Pooling
Integrated Voice and Data
Advanced Phone Features
Security Monitoring
Video Transmission
Interworking of Switches

The D channel distinguishes ISDN from today's telecommunications services. It serves as an umbilical cord between intelligent customer premises equipment and intelligent carrier-based switching systems. Using the D channel, customers will be able to control the use of B channels to support different applications on a call-by-call basis. One B channel could be used to access WATS during the day, then could be reconfigured to support access to a high-speed data service at night.

The D channel can be used to configure B channels in one common pipe. Six B channels could be used as a group to achieve a 384 kbit/s channel, an HO channel.

The full 1.54 Mbit/s capacity of a primary-rate interface can be obtained by using the D channel of a separate, parallel, primary-rate interface to provide the signaling information. A single primary-rate D channel can support signaling for up to 40 parallel links.

This flexibility enables customers to optimize various network services using a terminal or small computer to reconfigure the access channels in response to changing network conditions.

When ISDN becomes available, large business customers are likely to maintain primary and basic-rate interfaces to local-exchange carriers and primary-rate links to their long-haul carrier. The local carriers will provide the T-1 facilities to link customers to long-distance carriers. The circuits will not be switched; instead, they will only be wired through the local-exchange-carrier's switch offices.

The premises controller will support primary and basic-rate interfaces known *I reference points*. There are four reference-pointinterfaces (R,S,T, and U) defined by AT&T (see FIG. 8-5). These reference points define the network interface points between telephone company property and customer premises equipment. The R interface connects non-ISDN equipment to ISDN services. The other three reference points maintain separation between the signaling and bearer channels.

APPLICATIONS

Besides allowing customers to optimize their networks, ISDN opens up new applications. One example is telemarketing using automatic call distributors (ACD) to route incoming calls to waiting operators.

AT&T has found that ISDN can be used to spread traffic evenly over multiple ACD sites. The ACDs periodically relay their status to an 800 service database using D-channel access links. Incoming 800 toll-free calls can be routed based on the time of day and point of origin as well as by traffic load. AT&T found that the time it took calls to be answered was reduced by 75 percent, and the number of callers who hung up was reduced by an order of magnitude. This type of implementation also can cut costs by reducing the number of operators required.

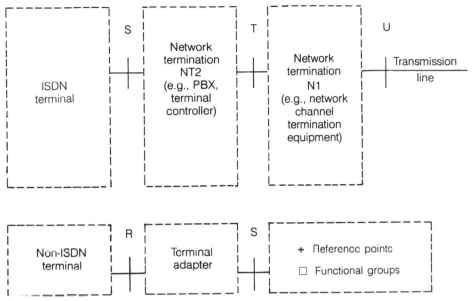

Fig. 8-5. ISDN reference-point interfaces.

The implementation of ISDN requires questions such as:

- What are the specific network requirements?
- What applications belong on private resources?
- What applications should be placed on public networks?
- How much integration of voice and data applications is needed?

The fundamental cost structure of primary-rate ISDN links will be similar to T-1 rates, but the cost of data messages carried on the signaling systems is another question. Access costs are established already. What is different is the interface in which you move messages across the signaling network.

Companies purchasing communications equipment such as PBXs should understand ISDN's role. A PBX installed under a six-year lease will still be expected to perform competitively after ISDN has become a significant force. Firms using longer lease or depreciation periods have a greater risk of obsolescence.

Effective communications-planning architectures in the '90s must specify corporate policies on many issues and technologies, including voice/data integration, standards, equipment-procurement procedures, and ISDN.

The answers to these questions will vary, depending on a corporation's business and management policies. Also, it will be necessary to consider the corporate requirements and formulate questions for vendors on ISDN, such as:

- How well does the vendor support standards?
- What is its migration plan toward the new standards or technologies?
- How will proprietary components and private networks work with open-systems protocols and public resources?

Whatever role a firm chooses for ISDN, that role will be an evolutionary step toward meeting requirements. Networks in the 1990s will be dynamic and flexible but generally not based completely on ISDN.

AN INGEGRATED SYSTEM
ARCHITECTURE FOR PRIVATE NETWORKS

An example of an integrated system architecture is Cyber Digital's MSX system, which is an integrated voice/data system that allows users to access data systems without the use of modems. Dissimilar asynchronous devices communicate with each other using automatic data flow control and protocol conversions. The MSX system can support asynchronous devices at 19.2 kbps and synchronous devices up to 64 kbps using a circuit-switched mode. Asynchronous packet-switched data is supported in the X.25 format. A star topology local-area network is employed as shown in FIG. 8-6.

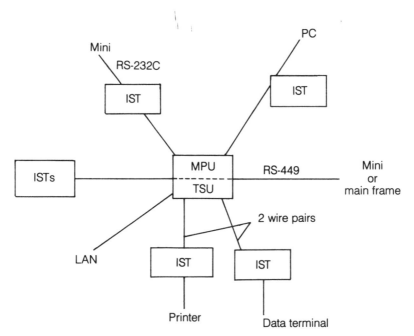

Fig. 8-6. An integrated system architecture.

A distributed microprocessor architecture is used with parallel processing. Six types of printed circuit boards are used:

1. A main processor unit (MPU)
2. A time switch unit (TSU)
3. A T1 interface unit (T1-U)
4. A voice interface unit (VIU)
5. A data voice unit (DVU)
6. An integrated service terminal (IST)

The MPU uses an 8086, 16-bit microprocessor and addresses 2 Mbytes of RAM. The basic software requires about 1 Mbyte of memory. The MPU controls the 12.8 Mbps pack/data bus and provides the required interface, real-time clock, error detection and correction, and an RS-C232 interface for a service terminal.

The TSU uses a Z-80, 8-bit microprocessor to control the 32.7 Mbps circuit bus providing 512 TDM 64-kbps circuits for trunks, voice, and data. The TSU also provides the system clock, tone generation, and digital conferencing.

The T1-U also uses a Z-80 and supports two 1.544 Mbps T1 D3 circuits that provide 24 or 48 digital trunks. The T1-U provides a direct T1 connection to digital-telephone central offices, fiberoptic, digital microwave, and digital satellite systems.

The T1-U provides one X.25 input/output port (DTE or DCE) with an RS-449 interface. The T1-U can be configured with software to use either 24 or 48 time slots, with 24 time slots per T1 trunk. The complete range of analog telephone central office trunking such as E&M, TIE, FX, and DID is supported using a D3 channel bank connected to one T1 circuit.

The VIU uses a Z-80 and provides a two-wire interface to standard analog telephone equipment. Each VIU supports 16 DTMF or rotary dialing telephones and requires 16 times slots.

The DVU uses a Z-80 and supports eight ISTs. Each DVU port uses time-division multiplexing to support data, voice, and control signaling. A codec is located in each IST for digital voice transmission in a PCM format at 64 kbps. Each DVU requires 16 time slots.

The IST is a type of digital telephone. Each IST has 12 or 20 pushbuttons and circuit-board options such as asynchronous or synchronous data boards with RS-232-C or RS-449 connectors. Options also include an LCD display. Each IST acts as an intelligent terminal with its own microprocessor and memory.

A Data Voice Unit (DVU) card can support eight ISTs. Digitized voice, data, text, graphics, control signaling, and power are all transmitted over standard two-wire twisted telephone pairs.

The software interface is based on user-friendly English-language commands with prompts and help functions. The system will continue to operate without service degradation when faults occur because of the use of redundant elements. Connection costs range from $200 to $700 per line, depending on the features and options needed.

The system is nonblocking and provides 36 CCs of traffic for trunks or standard analog telephones and 72 CCs of traffic for ISTs, 36 CC voice, and 36 CC data.

COMMUNICATIONS

The system provides asynchronous and synchronous data, digitized voice, text, graphics, and messaging using packet store and forward modes over standard two-pair twisted wires. HDLC (high-level data-link Control) and MDPSK (modified differential phase-shift keying) are employed. X.25 and T1 gateways are available. Three databuses are used for transmitting digital information.

1. **The Circuit Bus.** This is a 32.7 Mbps, 8-bit, time-division multiplexed bus with 512 time slots, of 64 kbps for voice and data.
2. **The Packet/Data Bus.** This is a 12.8 Mbps, 8-bit bidirectional bus that carries the control and signaling information and up to 64 kbps of packet-switched data in the X.25 format.
3. **A Shugart Associates Standard Interface (SASI) bus.** This is an 8-bit, parallel, 10-Mbps bus to connect to the disk storage devices.

Asynchronous data to or from an IST or ISU is routed using either the circuit or packet/data bus. An X.25 serial input/output port with an RS-449 interface is provided on each T1 interface unit.

The packet/data bus supports multiple X.25 ports using dynamic allocation of the total 64 kbps bandwidth. Data can be submultiplexed over T1 channels or other digital trunks in 8 kbps increments.

Bibliography

Abbateillo, J., and R. Sarch., eds. *Telecommunications and Data Communications Factbook*. New York: McGraw-Hill Bk. Co., 1987.

Abramson, N., and F.F. Kuo, eds. *Computer-Communications Networks*. Englewood Cliffs, N.J.: Prentice Hall, 1973.

Acampora, A.S., et al. "Performance of a Centralized-Bus, Local-Area Network." *Proceedings of Localnet 1983*. New York: Online Publications Ltd., Pinner, U.K., September 1983.

Advances in Cryptology: Proceedings of Crypto '82. New York: Plenum Press, 1983.

Alford, R.S. and C. Holmberg, "Implementing Automatic Route Selection in Networks." *Data Communications*, November 1987.

Allison, D.R. "A Design Philosophy for Microcomputer Architectures." *Computer*, 1977, 10,2.

Anderson, D.A. "Design of Self-Checking Digital Networks Using Code Techniques." PhD dissertation, Report R 527. Urbana, Il: University of Illinois, October 1971.

Andrews, M. *Principles of Firmware Engineering in Microprogram Control*. London: Computer Science Press, Inc., 1980.

AT&T. "The Extended Framing Format Interface Specification." Compatibility Bulletin no. 142. Parsippany, N.J., American Telephone & Telegraph-Communications, 1981.

Avizienis, A. "Fault-Tolerant Computing—An Overview." *Computer*, January-February 1971.

Bibliography

Axner, D.H. "Stat MUXes Are Alive and Well." TPT, June 1988.

Bauer, K. "LAN Security." Dallas, InfoLAN 88, April 27, 1988.

Baugh, K. "Intra-City Networking Via Cable TV." InfoLAN 88, April 27, 1988.

Baum, A., and D. Senzig. "Hardware Considerations in a Microcomputer Multiprocessor System." COMCON, San Francisco February 1975.

Bell, C., and A. Newell. *Computer Structures*. New York: McGraw-Hill, Bk. Co., 1970.

Bell, Trudy E. "Long-distance Fiberoptic Networks, Direct Broadcast Satellites, and Low-Cost PBXs Bring Increased Communications Capacity to Customers." *IEEE Spectrum*, January 1984.

Blahut, R.E. *Theory and Practice of Error Control Codes*. Reading, Mass.: Addison-Wesley, Pub. Co., Inc. 1983.

Bosworth, Bruce. *Codes, Ciphers, and Computers*. Rochelle Park, NJ: Hayden Books, 1982.

Bracker, W.E., and R. Sarch, *Cases in Network Design*. New York: McGraw-Hill Bk. Co., 1985.

Breuer, M.A., and A.D. Griedman. *"Diagnosis and Reliable Design of Digital Systems."* Woodland Hills, Cal.: *Computer Science* Press, 1976.

Brooks, F.P. "An Overview of Microcomputer Architecture and Software." Micro Architecture, EUROMICRO 1976 Proceedings.

Burnside, D.F. "Last-Mile Communications Alternatives." *TPT*, April 1988.

Burton, D.P. "Handle Microcomputer I/O Efficiently." *Electronic Design*, 13, June 21, 1978.

Burzio, G. "Operating Systems Enhance μCs." *Electronic Design*, June 21, 1978.

Carne, E.B. "New Dimensions in Telecommunications." *IEEE Comm. Magazine*, 20(1), January 1982.

CCITT Yellow Book, Vol. IV. 4. *Specifications of Measuring Equipment*. Geneva: ITU, 1981.

CCITT Yellow Book, Vol. II.2. *Telegraph and Telematic Services Terminal Equipment*. Geneva: ITU, 1981.

CCITT Yellow Book, Vol. VIII.1. *Data Communication Over the Telephone Network*. Geneva: ITU, 1981.

CCITT Yellow Book, Vol. III.3. *Digital Networks—Transmission Systems and Multiplexing Equipment*. Geneva: ITU, 1981.

Chandy, K.M., and M. Reiser. eds. *Computer Performance*. Amsterdam, Netherlands: North-Holland, 1977.

Charney, H. "The Future of Work Group Systems." InfoLAN 88, April 27, 1988.

Chu, Y. *Computer Organization and Microprogramming*. Englewood Cliffs, N.J.: Prentice Hall, 1972.

Clark, G.C., Jr., and J.B. Cain. *Error-Correction Coding for Digital Communications*. New York: Plenum Press, 1981.

Cooper, Edward. "CATV/Broadband Overview for Data and Telecommunications Managers." TR-81052. SYTEK, Inc., November 1981, copyright October 1981.

Crick, A. "Scheduling and Controlling I/O Operations." *Data Processing*, May-June 1974.

Crochiere, R.E., and J.L. Flanagan. "Current Perspectives in Digital Speech." *IEEE Communications* 21(1), January 1983.

Crossett, J.A. "Monitor and Control of Digital Transmission Systems." International Conference on Communications, 1981.

Dal Cin, M. "Performance Evaluation of Self-Diagnosing Multiprocessor Systems." Conference on Fault-Tolerant Computing, Toulouse, France, June 1978.

Data Transmission Over the Telephone Network. Yellow Book Vol. VIII.I CCITT. Geneva, 1981.

"Demultiplexing Considerations for Statistical Multiplexers." *IEEE Transactions on Communications*, Vol. COM-20, No. 3, June 1982.

Denning, Dorothy. *Cryptography and Data Security*. Reading, Mass.: Addison-Wesley Publishing Co., Inc. 1982.

Digital Equipment Corp., Intel Corp., and Xerox Corp. "The EtherNet: Data Link and Physical Layers." September 1980.

——————. *Introduction to Local-Area Networks*. 1982.

Dowsing, R.D. "Processor Management in a Multiprocessor System." *Electronic Letters*, November 1976.

Dugan, P. "The STARLAN Standard." Dallas, InfoLAN 88, April 27, 1988.

Duke, David A., and Donald B. Keck. "Single-Mode Fiberoptic Features." *Telecommunications*, Vol. 17, No. 12, December 1983.

Eastman, C., J. Lividini, and D. Stoker. "A Data base for Designing Large Physical Systems". NCC: AFIPS, 1975.

Eastman, C.M., and M. Henrion. "GLIDE: A Language for Designing Information Systems." SIGGRAPH '77 Proceedings, published as *Computer Graphics*, 11 (2), 1977.

Eckhouse, R.H., Jr. *Minicomputer Systems*. Englewood Cliffs, N.J.: Prentice Hall, 1975.

Eckman, D.P. *Automatic Process Control*. New York: John Wiley & Sons, Inc., 1958.

"E-Mail and LANs." New software, *Data Communications*, June 1987.

"E-Mail Boost." New software, *Data Communications*, October 1987.

"E-Mail Gateways." New software, *Data Communications*, August 1987.

"E-Mail Plus Disoss." New software, *Data Communications*, July 1987.

"E-Mail Plus." New software, *Data Communications*, July 1987.

Engel, S., and R. Granda. *Guidelines for Man/Display Interfaces*. IBM Technical Report TR 00.2720. Poughkeepsie, N.Y., 1975.

English, R.E. "Systems Management of a CAD/CAM Installation." CADCON East '84, June 13, Boston. New York: Morgan-Grampian.

Enslow, P.H., Jr., (Ed.) *Multiprocessors and Parallel Processing*. New York: John Wiley & Sons, Inc. 1974.

Farnbach, W.A. "Bring Up Your μP Bit-by-Bit." *Electronic Design*, 24, 15, July 19, 1976.

Faux, I.D., and M.J. Pratt. *"Computational Geometry for Design Manufacture."* New York: John Wiley & Sons, Inc., 1979.

Feher, Kamilo. *Digital Communications*. Englewood Cliffs, N.J.: Prentice Hall, 1981.

Feiner, S., S. Nagy, and A. Van Dam. "An Integrated System for Creating and Presenting Complex Computer-Based Documents." Proceedings 1981 SIGGRAPH Conference, published as *Computer Graphics*, 15 (3), August 1981.

Fields, A., R. Maisano, and C. Marshall. "A Comparative Analysis of Methods for Tactical Data Inputting." Army Research Institute, 1977.

Finch, B. "Voice and Data: The Role of the PBX in the Corporate Data Network." InfoLAN 88, April 27, 1988.

Foster, C.C. *Computer Architecture*. New York: Van Nostrand Reinhold Company, Inc., 1970.

Franklin, M.A., S.A. Kahn, and M.J. Stucki. "Design Issues in the Development of a Modular Multiprocessor Communications Network." Sixth Annual Symposium on Computer Architecture, April 23-25, 1979.

Freedman, M.D. Principles of Digital Computer Operation. New York: John Wiley & Sons, Inc., 1972.

Freeman, R.L. *Telecommunications Transmission Handbook*, 2nd ed. New York: John Wiley & Sons, Inc., 1981.

Friedman, A.D., and P.R. Memon. *Fault Detection in Digital Circuits*. Englewood Cliffs, N.J.: Prentice Hall, 1971.

Gardner, M. "LAN Standards." Dallas, InfoLAN 88, April 27, 1988.

Garland, II. *Introduction to Microprocessor System Design*. New York: McGraw-Hill Bk. Co., 1979.

Garrett, M.A. "Unified Non-Procedural Environment for Designing and Implementing Graphical Interfaces to Relational Data-Base Management Systems. Ph.D. dissertation, George Washington University, Washington, D.C., 1980.

Gear, C.W. *Computer Organization and Programming*. New York: McGraw-Hill Bk. Co., 1974.

Germann, J.J. "Using Special Processors to Enhance Engineering Workstations." CAD/CAM West '84, February 7-9, 1984, San Fransisco. New York: Morgan-Grampian.

Geyer, K.E., and K.R. Wilson. "Computing with Feeling." Proc. IEEE Conference on Computer Graphics, Pattern Recognition and Data Structure, May 1975.

Glasgal, Ralph. *Techniques in Data Communications*. Dedham, Mass.: Artech House, Inc., 1983.

Gloge, D.C., et al. "Characteristics and Operation of the TF4E-432Mb \ s repeater line." 1984 International Conference on communications.

Goldberg, Adele, and David Robson. A Metaphor for User Interface Design. Palo Alto, Cal.: Xerox Palo Alto Research Center, 1979.

Gonzalez, M.J., and C.V. Ramamoorthy. Parallel Task Execution in a Decentralized System. IEEE Transactions on Computers, December 1972.

Gooch, R.P. "DCEM: A DSI System for North America and Europe." 1982 International Conference on Communications.

Bibliography

Graaf, J. "The Role of Data PBXs in Networking." Dallas, InfoLAN '88, April 27, 1988.

Grimsdale, R.L., and D.M. Johnson, "A Modular Executive for Multiprocessor Systems." Sheffield, England: Trends in On-Line Computer Control Systems, April 1972.

Groff, G.K., and I.F. Muth. *Operations Management: Analysis for Decisions.* Homewood Ill.: Irwin, 1972.

Guedj, R., et al, ed. *Methodology of Interaction.* North-Holland, Amsterdam, 1980.

Hamilton, M., and S. Zeldin. "Higher Order Software—A Methodology for Defining Software." IEEE Transactions on Software engineering, SE-2, 1, March 1976.

Hanau, P.R., and D.R. Lenorovitz. "Prototyping and Simulation Tools for User/Computer Dialogue Design." SIGGRAPH '80 Proceedings, published as *Computer Graphics*, 14 (2), July 1980.

Hansen, W. "User Engineering Principles for Interactive Systems." Proceedings of Fall 1971, Joint Computer Conference.

Harris, J.A., and D.R. Smith. "Hierarchical Multiprocessor Organizations." Fourth Annual Symposium on Computer Architecture, March 23-25, 1977.

Hayes, P., E. Ball and R. Reddy. "Breaking the Man-Machine Communication Barrier." *Computer*, 14 (3), March 1981.

Herot, C.F., et al. "A Prototype Spatial Data-Base-Management System." SIGGRAPH '80 Proceedings, published as *Computer Graphics*, 14 (2), July 1980.

Held, G. and R., Sarch. *Data Communications: A Comprehensive Approach.* New York: McGraw-Hill Bk. Co., 1983.

Hill, F.J., and G.R. Peterson. *Digital Systems: Hardware Organization and Design.* New York: John Wiley & Sons, Inc., 1973.

Hooker, S. "Building Fault-Tolerant Networks." Dallas InfoLAN 88, April 27, 1988.

Hordeski, M.F. "Digital Control of Microprocessors." *Electronic Design*, December 6, 1975.

_____. "Innovative Design: Microprocessors." *Digital Design*, December 1976.

_____. "Balancing Microprocessor-Interface Tradeoffs." *Digital Design*, April 1977.

_____. "Future Microprocessor Software." *Digital Design*, August 1977.

_____. "Radiation and Stored Data" *Digital Design*, September 1977.

_____. "Microprocessor Chips." *Instrumentation Technology*, September 1977.

_____. "Fundamentals of Digital Control Loops." *Measurements & Control*, February 1978.

_____. "Using Microprocessors". *Measurements & Control*, June 1978.

_____. *Illustrated Dictionary of Microcomputer Terminology*. Blue Ridge Summit, Pa.: TAB BOOKS, Inc., 1978.

_____. *Microprocessor Cookbook*. Blue Ridge Summit, Pa.: TAB BOOKS, Inc., 1979.

_____. "Selecting Test Strategies for Microprocessor Systems." ATE Seminar Proceedings, Pasadena, Calif., January 1982. New York: Morgan-Grampian.

_____. "Selection of a Test Strategy for MPU Systems." *Electronics Test*, February 1982.

_____. "The Impact of 16-Bit Microprocessors." Las Vegas: Instrumentation Symposium Proceedings, May 1982 (Research Triangle Park, N.C. Instrument Society of America).

_____. "Diagnostic Strategies for Microprocessor Systems." ATE Seminar Proceedings, Anaheim, Ca., January 1983. New York: Morgan-Grampian.

_____. *Microprocessors in Industry.* New York: Van Nostrand Reinhold Co., Inc., 1984.

_____. *The Design of Microprocessor Sensor and Control Systems*. Reston, VA.: Reston, 1984.

_____. "CAD/CAM Equipment Reliability." Western Design Engineering Show and ASME Conference, San Francisco, December 5, 1984.

_____. "Specifying and Selecting CAD/CAM Equipment." CAD-CON West, Anaheim, January 14-17, 1985. New York: Morgan-Grampian.

_____. "A Tutorial on CIM/Factory Automation." Western Design Engineering Show and ASME Conference, Anaheim, December 12, 1985.

Bibliography

_____. *CAD/CAM Techniques*. Reston, Va.: Reston, 1986.

_____. *Microcomputer Design*. Reston, Va.: Reston, 1986.

Hornbuckle, G.D. "The Computer Graphics/User Interface." *IEEE Trans.* HFE-8 (1), March 1967.

Houtzel, A. "The Graphics Side of Group Technology. CADCON East '84, Boston, June 13. New York: Morgan-Grampian.

Hudry, J. "Man-Machine Interface Issues." *Computer Graphics World*, April 1984.

Intel Corp. *8086 User's Guide*. Santa Clara, Calif.: Intel, 1976.

Irani, K., and V. Wallace. "On Network Linguistics and the Conversational Design of Queueing Networks." *Journal of the ACM*, 18, October 1971.

Ishiguro, T., and K. Inuma. "Television Bandwidth Compression by Motion-Compensated Interframe Coding." *IEEE Communications*, 20(6), November 1982.

Ito, T., K. Nakagawa, and Y. Hakamada. "Design and Performances of the F-400M Trunk Transmission System Using a Single-Mode Fiber Cable." 1982 International Conference on Communications.

Jackson, M.A. *Principles of Program Design*. New York: Academic Press Inc., 1975.

Johnson, S., and M. Lesk. "Language Development Tools." The Bell System Technical Journal, 57 (6, 2) July-August 1978.

Jones, J.C. *Design Methods*. New York: Wiley-Interscience, 1970.

Kamae, T. "Public Facsimile Communication Network." *IEEE Communication* 20(2), March 1982.

Kaplan, M., and D. Greenberg. "Parallel Processing Techniques for Hidden-Surface Algorithms." SIGGRAPH '79 Proceedings, published as *Computer Graphics*, August 1979.

Katayama, S., M. Iwamoto, and J. Segawa, "Digital Radio-Relay Systems in NTT." 1984 International Communications Conference.

Kay, Alan C. "Microelectronics and the Personal Computer." *Scientific American* 237 (3), September 1977.

Kennedy, J.R. *A System for Timesharing Graphic Consoles*. FJCC. Washington, D.C.: Sparton Books, 1966.

Kilgour, A.C. "The Evolution of a Graphic System for Linked Computers." *Software—Practice and Experience*, 1, 1971.

Klingman, E.E. *Microprocessor Systems Design.* Englewood Cliffs, N.J.: Prentice Hall, 1977.

Klipec, B. "How to Avoid Noise Pickup on Wire and Cables." *Instruments & Control Systems*, December 1977.

Knapp, J.M. "The Ergonomic Millennium." *Computer Graphics World*, June 1983.

Knuth, D.E. *The Art of Computer Programming, Vol. 1: Fundamental Algorithms.* Reading, MA.: Addison-Wesley Publishing Co., Inc., 1973.

Kohonen, T. *Digital Circuits and Devices.* Englewood Cliffs, N.J.: Prentice Hall, 1972.

Konheim, A.G., and R.L. Pickholtz. "An Analysis of a Voice/Data Integrated Multiplexer." *IEEE Transactions on Communications*, January 1984.

Kostas, D.J. "Transition to ISDN—An Overview." *IEEE Communications*, Vol. 22, No. 1, January 1984.

Kriloff, H. "Human-Factor Considerations for Interactive Display Systems." In S. Trem, ed., Proceedings ACM/SIGGRAPH Workshop on User-Oriented Design of Interactive Graphics Systems, ACM 1976.

Kuck, D.J. *The Structure of Computers and Computations.* Vol. 1. New York: John Wiley & Sons, Inc., 1978.

Labie, J. "The Use of Fiberoptics in LANs." InfoLAN '88, April 27, 1988.

Lampson, Butler W. "Bravo Manual." In *Alto User's Handbook*, Xerox Palo Alto Research Center, Palo Alto, Calif., November 1978.

Leahy, W. "Data-Base Management—Automated Graphics Generation." CADCON East '84, June 12, Boston, MA. New York: Morgan-Grampian.

"Learning X.25." New software, *Data Communications*, March 1987.

Leininger, M. "Present and Future Developments in Robotic Applications." Automated Manufacturing, Greenville, S.C., March 19-22, 1984.

Leventhal, L.V. *Microprocessor Software, Hardware, Programming.* Englewood Cliffs, N.J.: Prentice Hall, 1978.

Lidinsky, W.P. "LANs, Internetworking and Public Networks." Proceedings of the Symposium on Local-Area Networks. Sydney, Australia, December 14-16, 1982.

"Linking MCI mail." New software, *Data Communications*, August 1987.

"Linking Packages." New software, *Data Communications*, November 1987.

"Linking Up to Boost the Electronic Office." *Business Week*, March 1983.

Local-Area Subnetworks: "A Performance Comparison." Proceedings of IFIP 6.4 International Workshop on Local-Area Networks.

London, H.S., and T.S. Giuffrida. "High-Speed Switched Digital Service." *IEEE Communication*, 21(2), March 1983.

London, H.S., and D.B. Menist. "A Description of the AT&T Video Teleconferencing System." 1981 National Telecommunications Conference.

Lorin, H. *Parallelism in Hardware and Software*. Englewood Cliffs, N.J.: Prentice Hall, 1972.

Madnick, S.F., and J.L. Donovan. *Operating Systems*. New York: McGraw-Hill Bk. Co., 1974.

"Managing Networks." New software, *Data Communications*, August 1987.

"Managing Networks." New software, *Data Communications*, October 1987.

"Managing Networks." New software, *Data Communications*, June 1987.

"Managing VAX." New software, *Data Communications*, September 1987.

"Managing X.25." New software, *Data Communications*, May 1987.

Martin, Donald P. *Microcomputer Design*. Chicago: Martin Research Ltd., 1975.

"Martin, J. *Design of Man-Computer Dialogues*. Englewood Cliffs, N.J.: Prentice Hall, 1973.

Martin, W.A. "Computer Input/Output of Mathematical Expressions." Second Symposium Symbolic Algebraic Manipulation, ACM, March 1971.

McCool, M. "Interfacing CAD/CAM with ATE." ATE West Proceedings, January 10-13, 1983.

McGlynn, D.R. *Microprocessors*. New York: John Wiley & Sons, Inc., 1976.

McKay, C.W. "An Approach to Distributing Intelligence Among a Network of Cooperative, Autonomous, and Functional Computing Clusters." Automated Manufacturing, Greenville, S.C., March 19-22, 1984.

McManigal, D., and D. Stevenson. "Architecture of the IBM 3277 Graphics Attachment." *IBM Systems Journal*, 19 (3), 1980.

McNamara, John E. *Technical Aspects of Data Communications*, 2nd ed. Bedford, Mass: Digital Equipment Corp., 1982.

McNichol, J., S. Barber, and F. Rivest. "Design and Application of the RD-4 and RD-6 Digital Radio Systems." 1984 International Conference on Communications.

Meditch, J.S. *Stochastic Optimal Linear Estimation and Control*. New York: McGraw-Hill Bk. Co., 1969.

Metcalfe, R.M., and D.R. Boggs. "ETHERNET: Distributed Packet Switching for Local Computer Networks." *Communications of the ACM*, 19 (7), July 1976.

Meyrowitz, N., and M. Moser. "BRUWIN: An Adaptable Design Strategy for Window Manager/Virtual Terminal Systems." Proceedings of the 8th Annual Symposium on Operating Systems Principles, (SIGOPS), Pacific Grove, Calif., December 1981.

Midwinter, J.E. "Development of High-Bit-Rate Monomode Fiber Systems in the United Kingdom." 1982 International Conference on Communications.

Miller, N. "Bus-Oriented Graphics Systems." *Computer Graphics World*, May 1983.

Miller, R.B. "Response Time in Man-Computer Conversational Transaction." 1968 FJCC, AFIPS Conference Proceedings, Vol. 33. Montvale, N.J.: AFIPS Press, 1968.

Mollenauer, J.F., "Networking for Greater Metropolitan Areas." *Data Communications*, February 1988.

Moran, T. "The Command Language Grammar: A Representation for the User Interface of Interactive Computer Systems." *International Journal of Man-Machine Studies*, 15, 1981.

Moss, D. "Multiprocessing Adds Muscle to μPs." *Electronic Design* II, May 24, 1978.

Motorola Semiconductor. *MC68000 Microprocessor User's Manual*. Austin, Tex.: Motorola, 1979.

Mouthaan. "Long-Wavelength Optical Transmission Systems in Europe." 1982 International Conference on Communications.

Murrill, P.W. *Automatic Control of Processes*. Scranton, Pa.: International Textbook Co., 1967.

Myers, G.J. *Reliable Software Through Composite Design*. New York: Petrocelli Charter, 1975.

Myers, G.J. *Advances in Computer Architecture*. New York: John Wiley & Sons, Inc., 1978.

"NetView and NetView/PC." New software, *Data Communications*, March 1987.

"NetView Gains with Fortune 1000." New software, *Data Communications*, September 1987.

"NetView Processing." New software, *Data Communications*, January 1987.

"NetView/PC Links." New software, *Data Communications*, August 1987.

"Network Control." New software, *Data Communications*, May 1987.

"Network Design." New software, *Data Communications*, May 1987.

Newman, W.M., and Sproull, R.F. *Principles of Interactive Computer Graphics*, *2nd ed.* New York: McGraw-Hill Bk. Co., 1979.

Newton, R.S. "An Exercise in Multiprocessor Operating-System Design. Agard Conference on Real-Time, Computer-Based Systems, Athens, Greece, NATO Advisory Group on Aerospace R & D, May 1974.

Nick, J.R. "Using Schottky Three-State Outputs in Bus-Organized Systems." *Electronic Design News*, 19 (23) December 5, 1974.

Nielson, D. "The Role of Radio in Local-Area Data Distribution." *Journal of Telecommunication Networks*, 1(1), Spring 1982.

Norton, F.J. "CADD, Human Relations, and the Management Process." CAD-CON West '84, San Francisco, February 7-9. Morgan-Grampian: New York, 1984.

Nunn, M. "CAE/CAD/CAM Testability—An Overview of CADCON West '84'', San Francisco, February 7-9. New York: Morgan-Grampian, 1984.

Obenzinger, Mark, M. "The Personal Computer Market: A Profile for Growth." New York: Lehman Brothers Kuhn Loeb Research, January 1983.

O'Brien, Michael T. "A Network Graphical Conferencing System." RAND Corporation, Santa Monica, CA., 1979 (N-1250-DARPA).

Oppenheim, A.V., and S. Schafer. *Digital Signal Processing.* Englewood Cliffs, N.J.: Prentice Hall, 1975.

Ottinger, L. "Using Robots in Flexible Manufacturing Cells/Facilities." Automated Manufacturing, Greenville, S.C. March 19-22, 1984.

Patel, J.H. "Processor-memory Interconnections for Multiprocessors." Proceedings of the Sixth Annual Symposium on Computer Architecture, April 23-25, 1979.

Patterson, D.A., and C.H. Seguin. "Design Considerations for Single-Chip Computers of the Future." IEEE Trans. Comp., Comp., C-29, February 1980.

"PC Connectivity." New software, *Data Communications*, June 1987.

"PCs Talking." New software, *Data Communications*, April 1987.

Pearson, D.J. "Graphics Workstation Intelligence." *Computer Graphics World*, January 1983.

Peuto, B.L., and L.J. Shustek. "Current Issues in the Architecture of Micro-processors." *Computer*, February 1977.

Pferd, W. "A New Boost with Automated Data Capture." *Computer Graphics World*, March 1983.

Pinto, J. "Artificial Intelligence and Robotics in the Future Factory." *Test & Measurement World*, December 1983.

Preiss, R. "Storage CRT Display Terminals: Evolution and Trends." *Computer*, 11 (11), November 1978.

Proakis, John G. *Digital Communications*. New York: McGraw-Hill Bk. Co., 1983.

"Protocol Analyzer Can Test X.25, SNA." New products, *Data Communications*, August 1987.

"Protocol Analyzer Has Five Built-In Interfaces." New products, *Data Communications*, February 1987.

"Protocol Analyzer Runs at Speeds of Up to 72 Kbit/s."New products, *Data Communications*, April 1987.

"Protocol Analyzer Series Designed for ISDN and T1." New products, *Data Communications*, March 1987.

"Protocol Analyzer Tests ISDN, T1, X.25." New products, *Data Communications*, May 1987.

Quereshi, Shahid. "Adaptive Equalization." *IEEE Communications*, March 1982.

Renfrow, N. "Tools for Facilities Management." CADCON East '84, June 12, Boston, Mass. New York: Morgan-Grampian.

Riker, D.C. "Digital Transmission in the MCI Network." *IEEE Communications*, 22(10), October 1984.

Rosenburg, R. "Network Trauma." *Data Communications*, July 1987.

Ryan, D.J. "Making Sense of Today's Image Communications Alternatives." *Data Communications*, April 1987.

Ryan, G.P. "Managing Network Performance with PCs." *TPT*, July 1988.

Sarch, R., ed. *Basic Guide to Data Communications*. New York: McGraw-Hill Bk. Co., 1985.

——————. *Data Network Design Strategies*. New York: McGraw-Hill Bk. Co., 1983.

——————. *Integrating Voice and Data*. New York: McGraw-Hill Bk. Co., 1987.

Sazegari, S.A. "Network Architects Plan Broadening of Future ISDN." *Data Communications*, July 1987.

Schaeffer, E.J., and T.J. Williams. "An Analysis of Fault Detection, Correction and Prevention in Industrial Computer Systems." Purdue Laboratory for Applied Industrial Control, Purdue University, October 1977.

"Securing X.25 Networks via DES Hardware." New products, *Data Communications*, October 1987.

"Security Lapses Under Attack by Worried Network Users". *Data Communications*, September 1987.

"Security Module." New software, *Data Communications*, May 1987.

"Security Scheme Has Few Takers." *Data Communications*, February 1987.

Seybold, J. "The Xerox Professional Workstation." *The Seybold Report*, 10 (16), April 1981.

Shepherd, M., Jr. "Distributed Computing Power: Opportunities and Challenges." National Computer Conference, 1977.

Sheridan, W. "How to Make and Use Null Modem Cables." *Data Communications*, November 1987.

Shneiderman, B. "Human-Factors Experiments in Designing Interactive Systems." *Computer*, 12 (12), December 1979.

Shoch, J.F., et al. "Evolution of the EtherNet Local Computer Network." *Computer*, August 1982.

"SNA Connections." New software, *Data Communications*, August 1987.

"SNA/SDLC Support." New software, *Data Communications*, August 1987.

"SNA to OSI." New software, *Data Communications*, August 1987.

"SNA, X.25 Control." New software, *Data Communications*, June 1987.

Sobel, H.S. *Introduction to Digital Computer Design*. Reading, MA: Addison-Wesley Publishing Co., Inc., 1970.

Socci, V. "Microprocessors in Distributed Graphics." *Computer Graphics World*, May 1983.

Soucek, B. *Microprocessors and Microcomputers*. New York: John Wiley & Sons, Inc., 1976.

Sproull, R.F., and E.L. Thomas. "A Network Graphics Protocol." *Computer Graphics, 8(3), Fall 1974.*

Stallings, W., "Is There an OSI Session Protocol in Your Future?" Data Communications, November 1987.

Stone, Harold. "Critical Load Factors in Two-Processor Distributed Systems." *IEEE Transactions on Software Engineering*, SE-4 (3), May 1978.

Stone, H.S. *Introduction to Computer Architecture*. New York: McGraw-Hill Bk. Co., 1975.

Stusser, D.I., and R.S. Passafaro. "Evaluating Telemanagement Systems." *TPT*, July 1988.

"T1 CSU Can Accept Frequency Timing from External Source." *New Products*, September 1987.

"T1 MUX Handles Up to 500 Voice Data, Video Channels." *New Products*, August 1987.

"T1 Processors Now Sport NetView Interface." *New Products*, September 1987.

"T1 Products Reach for the High End of the Market." *New Products*, March 1987.

Taffel, A.B. "Packet-Satellite Networks." *Data Communications*, November 1987.

Takikawa, K. "Simplified 6.3 Mbit/s Codec for Video Conferencing." *IEEE Trans. on Communications*. Vol. COM-29, December 1981.

Tanenbaum, Andrew S. *Computer Networks*. Englewood Cliffs, N.J.: Prentice Hall, 1981.

Tasar, O., and V. Tasar. "A Study of Intermittent Faults in Digital Computers." AFIPS Conference Proceedings, Vol. 46. Montvalue, N.J.: AFIPS Press, 1977.

Taylor, S.A. "Falling Prices Expand T1 Options." *TPT*, March 1988.

Texas Instruments. *The Microprocessor Handbook*. Houston, Tex.: Texas Instruments, 1975.

Thomas, T.B., and W.L. Arbuckle. "Multiprocessor Software: Two Approaches." Conference on the Use of Digital Computers in Process Control, Baton Rouge, La., February 1971.

Thompson, J.E. "European Collaboration on Picture-Coding Research for 2 Mbit/s Transmission." *IEEE Trans. on communications.* Vol. COM-29, No. 12, December 1981.

Thurber, K.J., and G.M. Masson. *Distributed Processor Communication Architecture.* Lexington, Mass.: Lexington, 1979.

Tippie, J.W., and J.E. Kulaga. "Design Considerations for a Multiprocessor-Based Data-Acquisition System." *IEEE Transactions on Nuclear Science,* August 1979.

Toong, H.D., and A. Gupta. "An Architectural Comparison of Contemporary 16-Bit Microprocessors." *IEEE Micro,* May 1981.

Turpin, J., and R. Sarch, ed. *Data Communications: Beyond the Basics.* New York: McGraw-Hill Bk. Co., 1986.

Ungaro, C., ed. *Applying Standards.* New York: McGraw-Hill Bk Co., 1988.

————————. *Linking Microcomputers.* New York: McGraw-Hill Bk. Co., 1988.

————————. *Networking Software.* New York: McGraw-Hill Bk Co., 1987.

Ungerboeck, G. "Channel Coding with Multilevel/Phase Signals." *IEEE Trans. on Information Theory,* Vol. IT-28, No. 1, January 1982.

Vacroux, A.G. "Explore Microcomputer I/O Capabilities." *Electronic Design,* May 10, 1975.

van Dam, A. "Some Implementation Issues Relating to Data Structures for Interactive Graphics." *International Journal of Computer and Information Sciences,* 1 (4), 1972.

van den Bos, J. "Definition and Use of Higher Level Graphics Input Tools. SIGGRAPH '78 Proceedings, published as *Computer Graphics,* 12 (3), August 1978.

Wallace, V.L. "The Semantics of Graphic Input Devices." Proceedings of the SIGGRAPH/SIGPLAN Conference on Graphics Languages, published as *Computer Graphics,* 10 (1), April 1976.

Warner, J.R. "Device-Independent Tool Systems." *Computer Graphics World,* February 1984.

Warnock, J. "The Display of Characters Using Grey-Level Sample Arrays." SIGGRAPH '80 Proceedings, published as *Computer Graphics*, 14 (3), July 1980.

Wegner, W., ed. *Research Directions in Software Technology*. Cambridge, Mass.: MIT Press, 1978.

Weisberg, D.E. "Performance and Productivity in CAD." *Computer Graphics World*, June 1983.

Weller, D., and R. Williams. "Graphic and Relational Data-Base Support for Problem Solving. SIGGRAPH '76 Proceedings, published as *Computer Graphics*, 10 (2), Summer 1976.

Wilde, D.J. *Optimum Seeking Methods*. Englewood Cliffs, N.J.: Prentice Hall, 1964.

Williams, R. "On the Application of Relational Data Structures in Computer Graphics." Proceedings of the 1974 IFIP Congress, North-Holland.

Wolf, J.K., A.M. Michelson and A.H. Levesque. "On the Probability of Undetected Errors for Linear Block codes." *IEEE Transaction on Communications*. Vol. COM-30, February 1982.

Wong, E., and K. Youssefi. "Decomposition—A Strategy for Query Processing." *ACM Transactions on Data-Base Systems*, September 1976.

Yellowlees, R.A. "Voice/Data Integration: Planning is Key." *TPT*, March 1988.

Yourdon, E., and L.L. Constantine. *Structured Design*. New York: Yourdon Pr., 1975.

Zaks, R. *Microprocessors*. Berkeley, CA: Sybex, Inc., 1979.

Zimmerman, H. H., and M.G. Sovereign. *Quantitative Models for Production Management*. Englewood Cliffs, N. J.: Prentice Hall, 1974.

Index

E

U

Ultrix, 190
unattended autodial, 16
unauthorized entry, 146
Ungermann-Bass, 49, 50, 54
uniform call-distribution (UCD), 229
Unisys, 181
usage/user profile, 35

V

V.21 standard, 19
V.22 standard, 20
V.23 standard, 20
V.24 standard, 5, 19, 22, 23, 29
V.27 standard, 20
VAX/VMS systems, 190
VAXcluster, 190-192
Vernam cipher, 153
Video Network Services (VNS), 271
video teleconferencing, 270-271
Vines, 101
virtual network management, AI for, 105-106
virtual protocols, 21, 24
Viterbi decoding, 139
vocoders, 248
voice communication, 6-7, 15
  baseband networks and, 50
  circuit switching and, 8
  storage and forwarding of, 42
voice compression, cellular radio and, 248-252

voice management model, 68-70, 94
voice-band data, 279
voice-excited predictive coding (VEPC), 250, 251
voice-messaging systems, 230-237
voltage- controlled oscillators (VCO), 14
VSAT, 271-274
VTAM, 90, 91, 96

W

Wang, 54, 55, 181
Western Union, 262, 263
wide area telephone service (WATS), 44
Winchester disks, shared data on, 29
wiretapping security methods, 149-150

X

X.12 standard, 196
X.25 packet assembler/disassembler, 44, 45
X.25 protocol, 50, 99, 113, 148, 261
X.400 standard, 182, 245, 267-268
X3.66 protocol, 172
Xerox, 49, 179
Xerox Network Systems (XNS), 78, 113
XNS protocol, 29

Z

ZapMail, 261
zero-phase proposal, networks development and, 30